ALL THE WORLD IS KIN

Some Special People: in Fur and Feathers

by Bernice Espy Hicks

Illustrations by
Laney

Naturegraph Publishers, Inc.
Happy Camp, California

Library of Congress Cataloging in Publication Data

Hicks, Bernice Espy, 1913-
All the world is kin.

1. Domestic animals—Biography. 2. Animals, Legends and stories of. 3. Hicks, Bernice Espy, 1913- I. Title.
SF76.5.H5 1982 818'.5403 82-2086
ISBN 0-87961-116-2 AACR2
ISBN 0-87961-117-0 (pbk.)

Naturegraph Publishers, Inc.
P. O. Box 1075
Happy Camp, California 96039

Acknowledgements

This book was born out of my daughter's urging me to recall the experiences of a lifetime spent in the company of a succession of animals. The biographies in these pages are immeasurably enhanced by the superb artistry of that same daughter in her loving portraits and cover scenes. To my son and his family I owe the dimension of a "trial-run" reading, which gave some evidence of the book's appeal to the very young as well as to adults.

During its entire journey to completion, this book has been blessed by loving encouragement as well as tangible assistance from my dear friend Margaret Koerner (Gobi's other mother). It has also benefited from the loyal enthusiasm of my life-long friend, Virginia Deal Williams, who has given unstinting effort to its promotion. In the same cause, my esteemed friend, Dr. E. S. Inocencio has contributed his marketing expertise. My indebtedness to the important roles played by other friends is implicit and explicit in the text, and they will recognize my gratitude in the reading.

I cannot adequately acknowledge the enthusiasm, the careful and expert judgment, the indispensable supportive role which my editor, David L. Moore, has so consistently given me.

I would like also to thank my friend, the well-known writer, Aileen Fisher, for technical advice generously given.

Finally, I am deeply appreciative of the privilege granted by the publishers, Holt, Rinehart and Winston, to quote in the Preface from one of the most beautiful of American classics, *The Outermost House,* by Henry Beston.

Though I don't know where to place it in order of importance, there is the pervasive influence, the debt owed by my very genes, to my forebears, both immediate and distant, who bequeathed me the abilities, and provided me the early environment which resulted in my life-long happy addiction to these other "fellow prisoners" on our earth.

Table of Contents

Preface

In his classic book, *The Outermost House,* Henry Beston wrote "... For the animal shall not be measured by man. In a world older and more complete than ours, they move finished and complete, gifted with extensions of the senses we have lost or never attained, living by voices we shall never hear. They are not brethren, they are not underlings; they are other nations, caught with ourselves in the net of life and time, fellow prisoners of the splendour and travail of the earth." The kinship which exists among and between all Life is inevitably and increasingly being revealed to us in all its awesome intricacy and scope. It might even be, if we drew closer to these other creatures, fully recognizing this close relationship, we humans would find ourselves better able to come to grips with our own pressing inner problems.

The chapters in this book are the life histories of certain "people". They portray, as I witnessed them, and as my family and I played our parts in them, the gladnesses and the sadnesses, and the daily happenings which made up the lives of these animals. I have tried to bring to paper the qualities of character with which these remarkable individuals met life. There is an unavoidable interweaving with the human world throughout, but it was my intention to include only so much of our own species as was necessary to make the story of each animal more adequate. Inevitably, my personal perceptions and biases show through.

The animals who have shared their lives with me are mostly those man has chosen to bring into his web of existence, whether as companions, amusements, fellow workers, or as food supply. In each case, they adapted, extracting what knowledge of us they found necessary to make their lives not only tolerable but as much to their liking as possible. I suspect they often knew us better and more precisely than we ever did them.

Because my close to seventy years of living with and loving animals have both deepened and widened my exhilarating sense of belonging to a larger universe; because they have taught me how precious is simple, spontaneous joy of the existential moment; because they have shown me the quiet dignity of unpretentious honesty—in return for these and other gifts which they brought to my life, I could not let their importance, their right simply to have been what they were, drift into nothingness. All forms of life carry within mysterious legacies bequeathed to us from eons past. We in turn must pass on the best of what makes our individuality significant. Only then can we stand erect, worthy of belonging to the stream of life of which we are an indivisible part.

The corollary to a more accurate recognition of our place in the web of life is that these other forms must be granted their inalienable right to exist in dignity, their right to individuality recognized and respected. Since my "pets" have left me so much in their debt, this book becomes a plea to my fellow man, to make more than a condescending place in their hearts for these fellow beings with whom we walk on this small and beautiful planet.

Animals are such charming egoists! They operate in the confidence that they were born to be themselves, not to seem otherwise than they are. But most pets and other animals, living under the domination of humankind, have little ability or opportunity to create their own reality, because their treatment, unfortunately, is characterized less by partnership than by control.

When a human takes a "pet," he or she assumes a responsibility for that animal of more serious and far reaching proportions than many people seem willing to accept. Mere room and board are not enough. The animal is *not* merely a plaything, but equally a product of Nature, a fellow creature; one, moreover, whose welfare is placed in our hands by our choice of exercising stewardship. I use the word "stewardship" advisedly, since I believe, in the last analysis we can "own" nothing but ourselves. In the long term, perhaps, these fellow creatures may be more important to the total scheme of things than our species of hominid, only the latest to come upon the scene—and apparently in some danger of erasing ourselves and all other forms of life from that scene.

If this is blasphemy, so be it. I prefer to regard it as reverence for all life. To assume that intelligence, loving, joy and sorrow, laughter and grief are exclusively human characteristics and abilities—*this* to me is the most arrogant blasphemy, and an affront to the real, ongoing miracles constantly and everywhere about us. Western man needs to reappraise his flattering portrait of himself as a special creation who was given the unique privilege to exploit the world, its creatures and all its other resources. *Was* all this made for us to have dominion over? A species which does not even yet understand itself?

I am sure, with a conviction beyond my capacity to verbalize, that the tangible, physical thrill which flows through me when I achieve communication and understanding with another form of life is actually that rare privilege of glimpsing Unity, of even touching reverently something of the wonder-filled essence of Nature that is beyond my poor language to abstract. I knew it when little wild dove, Pecky, expressed her trust and her affection by choosing to settle on my neck for a nap; when Cindy knew she was dying, and gravely held out her paw to say "goodbye"; when Pitty Sing brought to *me* her first-born calf, whom she had hidden from everyone else; when so many times I would look, without saying anything, at Mimer sitting in his corner of the kitchen, and he would answer with a low "h-hm," not even opening his bill, his bright, dark eyes full of comprehension, his satisfaction in that communion reflected by a little shiver and the slightest flip of his wings. These other forms of life transcend our language, but we *can* communicate. There is truly a pre- and a post-language intuition by which they and we can know and share so much.

Many people don't have time, or don't think they do, to spend hours watching and learning from the quiet things a duck, or a cow, or a turtle does and is. They don't even think there could be fun in such an occupation. Obviously these biographies are not for them. But for those who do find a rich contentment in studying and loving other forms of life, I hope my words will bring alive to just a small extent the personalities of these "other nations," reinforce the fact that they do have characters, and finally, the right to develop their potentials.

THE AUTHOR, Bernice Espy Hicks, was born in Denver to a close-knit and practical family, the type that exhibited the best of true American pioneer character. Although Hicks received a degree from the University of Denver in philosophy and has served outstandingly in numerous state and local organizations, her deepest lifetime contribution is, perhaps, her genuine care and ability in sharing communication both with people and the many animals she has known.

Virginia

Chapter One

VIRGINIA

There is a folk saying which proclaims that as the twig is bent so grows the tree. The earliest twig I can remember helped to bend my tree permanently toward a yearning to understand and communicate with creatures other than my own kind. My paternal grandfather who possessed amazing rapport with animals must have predisposed the twig, but my early experiences reinforced and encouraged that direction.

The first scene I can pluck from memory puts me flat on my stomach on the scrubbed bare boards of our kitchen floor, in the house where I was born in Denver. Propped on my elbows, chin in my cupped hands, I am totally absorbed in watching a dozen or so yellow baby chicks running about in an enclosed space behind the warm stove. Sixty years older, I could be happily absorbed in doing exactly the same thing. So I am forced to believe that the twig helped to prepare the later tree!

A great part of the story of Virginia comes of necessity from reminiscences of my mother, as I was only four or so at the time and I retained only the few facts which were important to me. The kitchen stove I do remember. It was a black, cast-iron coal range set into the floor about a foot-and-a-half away from the wall. My mother and grandmother kept its nickel trim shining, as well as the general surface which they frequently daubed with a substance appropriately termed "lamp black" to produce a satiny finish. The coal bucket that stood beside the range, filled with its odd-shaped, shiny black chunks apparently held great charm for me. So my mother, being the soul of neatness, no doubt welcomed the diversion which Dad provided in my fourth spring in the form of a dozen baby chicks. They were not purchased for my amusement, but for the very utilitarian objective of eggs and meat. They were Rhode Island Reds, though that also did not matter to me then.

They were enchantingly, delightfully alive, and warm, and cuddly; their loud and mournful cheeps, if they stood irresolutely in the middle of their enclosure, changed like magic to soft, contented chirps in my hand. These baby chicks were housed, literally, in our kitchen where, as in households for generations past puppies, piglets, kittens, and all manner of small domesticated animals have known the enveloping warmth that radiated from the old coal stove. The distance allowed between the wall and the back of the stove was primarily a precaution against fire but was utilized in addition as an ideal enclosure for such infants. It kept them warm, out of the way, and easily contained.

We managed the fencing in by laying the extension leaves from our dining room table across the two ends. The ever-useful newspapers were spread on the floor, and fine sand (which my mother, who had just learned about germs, sterilized each morning in the oven) was scattered over the newsprint. A Mason jar filled with water, turned upside down, and resting on a pan, provided water, and a second like contraption served finely ground grains. In addition there was riced, hardboiled egg every morning which they loved and scrapped over more than they did over their grain. They would all pile over each other, cheeping, scrambling, pushing, pecking each other to get at the egg when my mother set it down in the pen. Usually everyone managed to get some, but I began to feel sorry for one chick, slightly smaller than the rest, who always got shoved out and stood teetering on the fringe, never quite daring to push, and indeed fearing the sharp pecks from the more aggressive ones. Each day this developing order of the chicken hierarchy became more rigid, and the lonely, tiniest chick failed, of course, to grow as fast as the other little rowdies.

My sympathy and loyalties were drawn to this little underdog, and I was allowed to feed it alone outside the pen, but the status quo had been established. Discrimination was deliberate and unremitting, and more and more this little one on the bottom of the pecking order became a sort of outcast, forced to sleep at the colder, outside edge of the huddled mass of yellow fuzz that slept on top of each other in jumbled chaos, and allowed to eat only after everyone else was finished. It was no wonder she became dependent on outside help, and also no wonder that she became quite used to me and to my handling her.

Why or when I decided to give her the fancy name of Virginia I do not know. It is one of the questions I forgot to inquire about from my parents, whose eternal silence now means that I can never share many such bits of information along with numerous facts of much more importance.

Chickens mature rapidly. In fact, poultry raisers have bred those birds whose early maturity means quicker profits. So I recall little more of Virginia's youth. She emerges most clearly from memory, therefore, as a full grown hen and my first companion. She remained always small for her breed, weighing only five pounds as against the average six-and-a-half of Rhode Island Red hens.

It must be a permanent residue of my childhood experience which causes chickens and their history to continue to be of great interest to me, both as individuals and as breeds. The homespun title of Rhode Island Red gives no hint of the romance and adventure which was involved when Yankee sea captains carried home to the Atlantic seaboard, in the mid-19th Century, exotic and beautifully burnished red cocks from the South Pacific. The beauty of the East was then mated to the practicality of the West, to hens who were Mayflower stock from Northern Europe. The result, a half century and many chicken generations later, the Rhode Island Red was on its way to being one of the most popular breeds in the United States. The cocks were a dark, rich red with long, generously arched black tails, of a beautiful bottle-green sheen, and with golden glints in their iridescent necks. Together with their hens of a lighter reddish brown they made a handsome flock; but it was even more pleasing to Yankee pragmatism that they were fast growers, good layers, and were economical to raise and keep. In 1909 a Rhode Island Red cost 80 to 90 cents *a year* to feed! I thank my birth date for putting me into that era when even city people could and did raise chickens in the backyard.

Virginia had a long, varied, and illustrious heritage. But to a four-year-old child in 1917, her origins were not nearly so important as her presence and her character then and there, that out of a brood of young and lusty chickens she was the smallest and not lusty at all. She needed to be encouraged and loved; she seemed to understand my feelings and to respond rather than to resist my wanting to hold her in my arms. She was the most delightful, satisfying, and cooperative companion of my days and when the other chickens would not let her eat with them I was happy that she came to the back door for me to feed her. Perhaps in time Virginia even ceased to feel herself a chicken, having been ostracized, as it were, from chicken society. For she certainly came to prefer my human company to that of her own kind and trusted me more than she did her fellow chickens. She was my playmate and the happiest association I knew in my early childhood.

Virginia entered into whatever games I devised. The most elaborate one I remember was a child's microcosm mirroring the main adult preoccupation of the day—World War I, which the United States had just entered. My mother had two cousins whose war participation was a topic of conversation and concern at our house. Cousin Birdie, who had volunteered to work at the Red

Cross Headquarters in Denver, deposited Kelly, her big black bulldog, at our house each day for babysitting. Since Kelly found his amusement in following me about, I acquired a second playmate. He must have been a considerably gentle bulldog to establish immediate rapport with and to be accepted as one of a congenial trio comprised of himself, a four-and-a-half-year-old child and a small Rhode Island Red hen. So Kelly joined our games, and Virginia, to her great credit, did not object. My most important game included both an unprotesting Virginia and an agreeable and cooperative Kelly. The game was based on the contribution my mother's other cousin was making to the war.

This cousin was unmarried, and very beautiful, and I worshipped Cheta with a sort of wordless admiration for as long as I can remember. She had gone abroad with the first contingent of Red Cross nurses. Two of my father's large red bandanas somehow made me into a Red Cross nurse just like Cheta. One became a scarf over my head and one wrapped around me became a skirt. My small red wagon was Cousin Cheta's ambulance and Virginia was the occupant of the ambulance—the unwitting victim of whatever war catastrophe I invented which required transporting her to the hospital. According to Cheta's letters home, which were read aloud and much discussed by all the family, there was always a column of ambulances and related gear as well as soldiers, filing slowly along the country roads of France. Therefore, Kelly became either the following string of vehicles or the army. He played his part with even more enthusiasm than Virginia, and in retrospect I think she was quite the braver of the two. For no other chicken I ever had subsequently would have permitted herself to be jerked and jounced along over boardwalks, riding alone and unprotected in an open wagon, followed by an enormous black monster, who panted and drooled right over one's tail, so to speak!

But so completely did Virginia trust me and so eagerly did Kelly join forces that I was able to play out the whole game to my complete satisfaction: the romantic Red Cross nurse pulling her wounded patients (Virginia) through the countryside, round and round the house, followed by the whole French army apparatus in the person of Kelly.

Every day, morning and night, when my grandmother went out to the chicken yard to scatter the grain for the chickens, Virginia would leave the flock and head for the back door. There she would stand waiting for me to come out the screen door with her food. Or, if she did not come immediately, I would go out with her food and call, and she would always come running, her slender body rocking from side to side, clucking eagerly to me the while. It was an unfailing pleasure to feel her bill gently pecking the grain I held in my hand for her.

Then came a day when Virginia was not at the back door waiting for me. I went out with my small can of grain, but Virginia did not come when I called. Neither could I find her out with the others, not in the chicken yard, not in the chicken house, nor in any other part of the backyard.

Trying to find a reason for this strange state of affairs, and never doubting but that her absence was only a temporary thing, I lit upon comments I had heard my parents and my grandmother make about changes that occurred in a chicken's laying habits, and that sometimes their disposition even altered when they underwent a particularly severe moult. So I came to the conclusion that Virginia must be moulting, and that it had so changed her character and her appearance I could not recognize her, or she me. I reported this conviction to my mother. And she reinforced my belief that this must indeed be so.

Years later I learned it was not a change of personality at all, nor a seasonal moult which had so affected Virginia. In fact, I suspect now that the very idea of her having changed so drastically was suggested in order to explain what my parents could not bear to reveal to my trusting mind. Hindsight convinces me it must have been sheer stupidity on my part that I did not probe more sharply into Virginia's sudden disappearance.

By habit and from necessity in those days most people who kept chickens made use of them not ony for their eggs, but for their meat. Most, if not all, of the young roosters out of a spring hatch were so dispatched. The pullets were kept because with time they would be earning their keep by laying. Eventually, when they became old they were seldom pensioned, but met instead a quick end as a chicken stew.

It seems that my father had gone one night into the chicken shed and reached on the roost for the nearest fowl he could put his hands on. One did this as quickly as possible, because the longer an intruder messed around in the chicken house at night the more frightened the chickens became, and pandemonium would commence with much squawking and flapping about in the dark.

When my father brought in the chicken he had just killed, and he and my mother looked at the very small body, limp and ready for plucking, my parents realized what had happened.

Chub

Chapter Two

CHUB

As a child I loved Chub for the individual he was and his willingness to be my friend. As an adult I love him additionally for the priceless inheritance I know he left me: the experience of his gentle disposition, his capacity for affection, his willingness to indulge my wishes, all of which stamped a large part of my future. In another sense also I can now better appreciate him and what he represented: a draft horse of no mean accomplishments, without whose like the Europeans who came to North America would never have been able to plough the fields, explore and occupy this continent, nor haul the goods back and forth which kept in business the towns, the farms, the mines all over the land.

Even more than the human sweat which has gone into the building of the United States of America, the sweat and muscle of millions of draft horses like Chubby provided literally the horse power that made everything else possible. The oxen, the burro, and his half-brother and half-sister the mule, all contributed their strength and intelligence, but it was the draft horse which performed most satisfactorily the greatest amount of all-round labor.

Gone is the last generation of men who enjoyed the intimacy which came with the care and feeding and use of the draft horse, the small daily pleasures of rapport with these intelligent and cooperative animals. This was my grandfather's generation; the transition to machines, from ice-wagons to trucks, from the hay wagon to the automatic baler, occurred during my father's era. Thus it was that my first companions in the wonderful world of horse flesh were the "pensioned" ice-wagon horses whom my father and grandfather took to the Ice Plant in the mountains west of Denver.

They had bought in 1913 a tract of land west of the very small town of Rollinsville, along South Boulder Creek and the route of the Denver and Salt Lake Railway, where they engaged in what was known as the Natural Ice Business. They constructed a dam, made a lake, cut the ice off it in winter, stored it in an ice house just to the east of the lake, and shipped it out over the railroad in the summer. In subsequent years Dad enlarged the entire ice operation and bought more contiguous land on both sides of the canyon, to

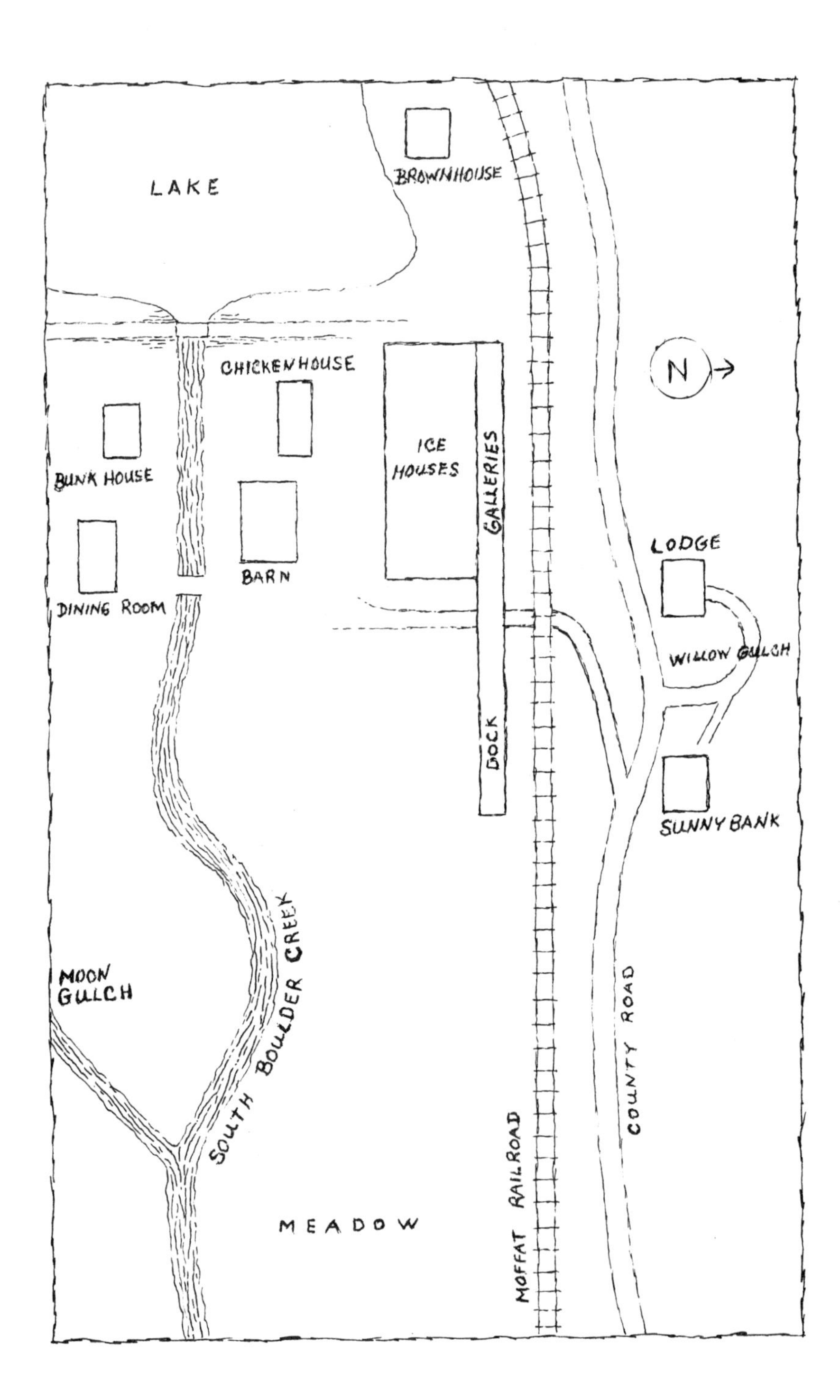
LAKE
BROWNHOUSE
CHICKENHOUSE
BUNK HOUSE
ICE HOUSES
GALLERIES
N
LODGE
BARN
DINING ROOM
WILLOW GULCH
DOCK
SUNNY BANK
SOUTH BOULDER CREEK
MOON GULCH
COUNTY ROAD
MOFFAT RAILROAD
MEADOW

encompass a small ranching operation. There the horses occasionally did farm work, and seasonally helped in the operations of the ice business.

Of the retirees, Chub was to me by far the most outstanding character. So was he also to my grandfather, a teamster in the original sense of the word, earning his living by driving his team for hire. Being a poor man financially, Bampa was unable to indulge in the blooded aristocracy of draft animals. He could afford only to buy what were then called "grade" horses, cross-breeds being reproduced as fast as possible by any and all matches of likely and unlikely characters, just so they had four feet and muscle and could be made to pull the loads of a hectic society "on the make."

Chub was *sui generis.* Whatever strains of however many heterogeneous backgrounds came together in Chub, they were most fortunately combined. He was the first horse I was to know well, and I have never since known one who topped him in intelligence, engaging charm, and good common horse sense. These were the traits for which he had been famous in his day. When I knew him, from the time I was eight or ten years old, Chubby was already an old man, retired and pensioned. So far as his alert brain, his interest and participation in his surroundings were concerned, however, he never retired.

Chub was just what his name indicated, a short-legged, chubby gelding. Small for a draft horse, the smallest by a couple of hundred pounds of any of the pensioners at the Ice Plant, small even for the dominant breed in his genes, the Percheron, he nevertheless had the agility, the clean cut, medium sized head, the short neck and trouble-free legs and feet, the intelligence and the initiative which aficionados have always claimed for the European/French Percheron. He was a bright bay, with a long, wavy black tail and long, silky black fetlocks. His mane and forelock, though, were short, stubby and wiry, almost as if they had been frizzed with a dry iron. His short forelock plus a small white star in the middle of his forehead somehow achieved a look of impertinence. But in his gentle, clear brown eyes shone his most prominent characteristics: intelligence, willingness, an eager curiosity which an unknown original trainer had preserved by a combination of thorough guidance and sympathetic understanding. As his age increased, the white star became somewhat indistinct in the general greying, and white hairs about his mouth gave him a grizzled appearance. Chub's ears were another of his distinctive features. They were the softest I have ever known to belong to a horse.

Bampa bought him at an auction held in the old stockyards of Denver, having served how many masters before, no one could ever know. He had already made his reputation at the stockyards as the most reliable team-breaker in the profession. The wildest, unbroken, mustang recruit coupled in harness with this small bay gelding could throw himself against the tongue, rear back, kick the traces, get himself tangled in his harness, but Chub never flinched nor bolted nor panicked. He was simply unflappable. When the driver called his order to go, Chub would move out slowly and cautiously acquainting his unruly teammate with the feel of the harness and the rub against his muscles, the hard bit in his mouth, the reins that suddenly pulled his head back or sideways. His steady, quiet example soon had the frightened and bewildered novice calmed down, and eventually more willing to put his shoulders against the breast board, to feel the belly band, the heavy collar over his neck, all of it moving, pressing, pulling against his sensitive flesh and against his will. Not many animals could be found to train successfully the unbroken horses who arrived constantly at the stockyards.

Since he was so highly regarded by the professional horse-traders, Chubby would have been held for a much higher price than Bampa could have afforded, except for the fact that the previous owner had demanded too much of the little bay. Chub fitted a description of the famous Justin Morgan breed: whatever he was hitched to generally had to give the first time. He had given all the strength he had and more to move a load, and in so doing had strained a shoulder. Like all slave flesh, he was auctioned at a discount because of his injury and disability. My grandfather, however, was a sharply discerning lover of good horse flesh, even one under a handicap. He reasoned in Chubby's case that he could be nursed back to at least near-normal. And that was how Chub became an ice-wagon horse.

In his new occupation Chub gained a measure of fame for himself and the ice-wagon he pulled, and made an equally indelible impression on an entirely different group of people. It all came about because of his ears, those ears that were as soft and pliable as any dog's. He was usually teamed with Tuck, who was old and almost white, and a half-hand higher than Chub, with a result not exactly smart nor fashionable in appearance. Tuck was a phlegmatic individual, pleased to have Chub do the thinking for both of them. He was also pleased to let Chub take the major share of a load, though he did not shirk if the driver were alert and prodded him. He also, by the same token, didn't mind Chub's using him to achieve what he wanted, provided it didn't require any effort on Tuck's part.

Chub did not like a bridle. He did not remonstrate against having it put on, but because of those soft ears he was able to work it off his head in a matter of moments, by rubbing his head against his teammate's shoulder. It did not matter where or when or for how long or how short a time, if left to stand in harness, Chub would be standing, calmly bridleless, when his driver returned to the wagon. The throat latch could still be tight under his jaws, but the bit would be out of his mouth, and the whole apparatus of straps and blinders dangling below his head in a most discomposed and unorthodox manner, with only a quizzical expression on his face to indicate his pleasure with what he had accomplished.

Since the day of an ice-wagon horse was made up of nothing but stops and starts and short absences of his driver, the teamster had to learn that this was a harmless idiosyncrasy of Chub's and not an alarming possibility for an accidental runaway, a phenomenon both frequent and frightening in the days when nothing moved without the power of horse flesh. Putting the bridle back on Chub's head after every stop became something more than a nuisance, and the driver soon found that he might as well leave Chub to his own inclinations, inasmuch as the horse knew even better than the teamster where each stop was. Moreover, he never in all his life spent with my family gave evidence that he entertained the thought of running away while in harness. So the driver allowed the bridle to hang loose under Chub's head, and merely put it back on at the end of the day to go across town to the barn. But outside Chub's circle of acquaintances along his ice route, who considered him something of a legend, a society that ran on horse power considered it neither proper nor safe for a working animal to go about with his uniform half off, so to speak.

All sorts of new possibilities in an exciting new life opened up to Chub and his fellow pensioners when they were moved to the Ice Plant to live out the rest of their days. And Chub quickly discovered new dimensions in his capacity for leadership, qualities which had never had much opportunity to develop so long as he pulled a wagon all day almost every day of the year, and spent his free time in a barn stall or adjacent yard. Now where he led, the others seemed unquestioningly to follow.

He found that fences had a tendency to holes, or at least places where they leaned pleasantly low, and that the temptation to step outside and explore what lay beyond was irresistible. Thus fairly early into his retirement Chub engineered his most spectacular escapade, one which took his buddies, Tuck and Tom and Floss, and the new riding mare Patsy and her filly, Emmy, far afield. For distance covered and duration of the adventure, it was by far his most resounding success. He managed to disappear almost every succeeding summer, but he never got as far nor kept the whole herd out for as many weeks. The weather had turned hot unseasonably early that summer, and the flies and mosquitoes were a torment to everyone. Chub somehow found a hole in the west end of the fence where it crossed the creek, and he just plodded through and kept on going, upward and westward.

Some seven miles up the canyon from the Ice Plant rises the Main Range, which forms the Continental Divide. Just before reaching its base, the old Denver and Salt Lake Railway veered to the north, and by a tortuous 12- to 15-mile climb, over a 4% grade, made its way to the top of the Divide, and wound another 12 miles down the other side. At the top was a mile or so of snow sheds erected against the Arctic rigors of the winter winds and snow at 11,000 feet. Since this route was one of the most gorgeous summertime scenic railroad trips in the entire country, a hotel was built on the very top of the Divide, at Corona, to accommodate the tourists who came in increasing numbers. Not surprisingly, the hotel did a very good restaurant business.

After some weeks and numerous fruitless searches of the area surrounding our property, my father began to hear stories from train crews of the D & S L, about a small motley herd of horses who hung around the snow sheds and along the crest of the Divide. The men would laugh about the scramble these horses made for garbage thrown out by the dining car chefs, by the section crews (maintenance men who lived in the accommodations inside the sheds on the Pass), and from the hotel dining room a hundred yards away. Dad inquired about the number of horses and their appearance. The men described first and most fully the one they took to be the leader. He was very friendly, they said, a modest-sized bay gelding, with a stubby forelock and mane, and a long black tail. He was, moreover, the one who pried off the tops of garbage cans and even opened screen doors that did not fit too snugly.

Though the appetite for garbage and the ingenuity at opening doors had never been in the catalogue of our horses' accomplishments, their descriptions fitted too neatly to be ignored, especially the one that sounded like Chub. Dad took a bridle and hopped the next freight going west. At Corona, he found waiting for their daily handout, Chub and Tuck, Floss and Tom, Patsy and Emmy and a burro they had picked up along the way. They were, everyone of them, fat and sleek, having spent the summer above timberline, away from the mosquitoes, the deer-flies and horse-flies which had pestered them down in the valley. They had fed well on the rich, nourishing tundra grasses plus the amplitude of garbage they had learned to eat. They had indeed developed a taste, which never left any of them so long as they lived, for table scraps. They had, in short, enjoyed themselves to the fullest, and when Dad approached them they asserted their new-found independence by being spooky and whirling away.

That is, all but Chub. As though he had been standing in the stable yard, he remained perfectly still while Dad talked to him and walked up to him. He offered no resistance to the bridle and seemed actually to enjoy having his nose stroked and a familiar hand run under his stubby mane. Having captured the leader there was no question but that the others would follow, even though they had refused to allow my father to get near them. So Dad mounted Chub, the herd strung along behind, and without further incident everyone finally got home. Dad had gambled on Chub's dependable character, on his unfailing cooperation. Thus he could bring back the entire bunch, some 15 miles and a very long day's ride, armed only with one bridle and his trust in Chub.

The taste acquired for garbage during that summer on the Divide led to interesting developments at the Ice Plant. The tricks Chub had learned about opening doors he not only did not forget but taught to every apt pupil who joined the herd so long as he lived. By working one side of his mobile upper lip, he could catch the protruding edge of the screen door on our dining room porch, four steps above ground, and swing it wide. Before the screen could swing back again Chub had inserted his whole neck clear to the shoulders. He absorbed the whack of the door on his side and proceeded to reach an incredible distance within. He was not above climbing a couple of steps to get further in if the commissary was stowed too far back. At various times he was discovered by the cook helping himself to the top layer of a crate of tomatoes, or lettuce, and one time he had sampled a round dozen out of a case of eggs.

He also learned most ingeniously how to eat corn off the cob. By bracing his very large round front feet at each end, he anchored the ear, and then, with his lips pulled back, he bit off the kernels as neatly as any of us could.

All was not leisure, however, and Chub's occasional periods of work embraced all the variety that a combination ranch and natural Ice Plant required. He pulled no more ice-wagons; he broke no more fractious youngsters to the harness; but he did pull the mower, the hay rake, the hay wagon, the snow plow to clear the ice on the lake for harvesting, and he was the most efficient power Dad ever had for operating the "johnny hook." This device was used when all the ice stored in the ice houses had been removed to the bottom layer or two below gallery level. It was composed of a long cable, with a very large hook on one end. The cable stretched from the back of the ice house out to the railroad spur in front, where it was guided around a big horizontal, flanged wheel, fastened onto the rails. The hook in the house was inserted at the back of a string of ice cakes, and Chub was at the other end of the cable outside. A smaller hook at that end caught into the singletree ring at the rear of his harness and it was Chub's job to walk beside the track, pulling the cakes up to the door of the Ice House. There a man stood on the gallery to swing them in front of a holdbar on the endless chain going by. The "johnny hook" operator in the Ice House would shout "Ready!" After his first step Chub judged the weight he had to pull, and how far he had to walk in order for the cakes to reach the door, even though he had his back to the whole operation and was some fifty feet away. Actually he almost needed no driver, for once through a task he could carry on on his own, just as he had on the ice routes in Denver. Nor during the entire "johnny hook" stint, did he need—or wear—a bridle!

Chub was even willing to become a riding horse! Before my father purchased the first of a string of riding horses, I was eager to ride horseback. In spite of obvious drawbacks as a riding mount, Chub was my inevitable choice, both because of his disposition and his size. My brother rode the taller Tuck, who was both unobliging and bored.

Though small for a draft horse, Chub was too broad, his head too big, his shoulders too bulging with muscle, his legs were too knobby at the knees and his feet very much too big and round. But because he was short enough for me, even as a child, to climb upon him, though my legs stuck out almost at right angles when I sat on his back, Chub became the first pleasant riding experience in my life. He was so gentle, so quick to learn, so cooperative that he responded almost immediately to neck reining, a complete reversal of what he had been taught all his life to understand, the direct pulling of the team reins.

I always preferred the riding bridle on him over the harness bridle, for its lesser bulk and straps then took nothing away from his stubby forelock and the irregular white star in the middle of his forehead. To me he looked the incorrigible, lovable pixy he was, free of the heavy collar and restraining bands of the wagon harness. But not much can be said affirmatively for the saddle, which did indeed sit oddly on his rather broad middle. His long wavy black tail which reached the ground, contrasting with the coarse and almost kinky, stubby forelock and mane, his knobby legs and big feet all gave away the truth that he just was not meant to be a svelt, sleek riding horse—but he tried his best. Many draft horses would never have allowed the familiarity of kids crawling over them, and would have resented anything more than harness straps on their backs since they were trained to pull, not carry. Chub never took to trotting, and even less to galloping, which all children like to do. On the few occasions when he did trot he had his own unique style, distinctly unfashionable, throwing his feet out to the side in a comical and exaggerated way. His fetlocks were so long they waved in the breeze he created. Whatever his shortcomings, this little bay gelding draft horse initiated me into the joys of horseback riding. The silent understanding between horse and rider walking together as one through a quiet wood, or stopping to enjoy a view, makes possible another whole dimension in the enjoyment of mountain solitude.

No one could know how old Chub lived to be. Perhaps his end was merciful. No matter how it came I would have mourned. The horses one mid-winter were standing on the lee side of some tall firs that grew near the creek, stolidly enduring the cold and terrifically strong winds that frequently reached velocities of one hundred miles per hour. Suddenly an unusually severe onslaught cracked the trunk of one of the largest of the trees, about twenty feet up, and it crashed to the ground. Chubby was standing in its path. . . .

I suppose there have been many "Chubs" in the annals of the draft horse. If there were some 23 million in the United States in the early 20th Century (as some accounts have estimated) there had to be some who were like him. But it is a sad commentary that we, as a people, never paused long enough in replacing the draft horse with our machines to do more than feed their no longer needed muscle to pet food manufacturers (as these accounts have maintained).

Considering their role in the building of these United States, they should have at least a statue erected somewhere, in bronze, dedicated to their contribution. But the only statues I have ever seen have been to human pioneers, who may or may not have a horse underneath or in front of them. The Icelanders have a memorial to the draft horse, which stands gracefully in the middle of a park in Reykjavik. If ever I am rich I think I shall commission one. And if I do so, I shall have the silent approval of many long-dead teamsters who also loved and admired the strength, the intelligence, the loyal service of their "Chubs."

Alice

Chapter Three

ALICE

During the year or so I had loved Virginia we lived next door to a family who had come from England, and their way with the English language delighted my ears. This was not their only charm: they also kept chickens.

The Withington's were the source of much joy for me long after we had moved away to another part of town and seldom saw them, because, after Virginia's mysterious disappearance they presented me with a replacement chicken out of their flock. A young pullet, she had been spared being consumed for Sunday dinner. I cannot remember why I was enamored of the name I chose for her, but I called her Alice.

On her Buff Orpington side she was the result of fifteen years of intensive breeding by a William Cook of Orpington, Kent, and so had a right to her pride because of the achievement she represented. Her other pedigree was the same as Virginia's Rhode Island Red. She carried her reinforced blue blood with an assurance and complacency fully as noticeable as was Virginia's apologetic timidity.

It seemed as though Alice combined the restrained manner of her English background with the hardiness of her new-world mixture. She was the quietest hen I ever had, so silent that it became an event when she spoke even a monosyllable. She ate calmly and noiselessly, unlike the other chickens, and clucked only when she had youngsters to control.

I thought we were fortunate because our house was on a lot whose size permitted only a small yard. It perforce brought the chicken yard very close to the house, so that I was able to watch from the window or to be with the chickens without my mother's being concerned because I was out of sight. Our proximity made the morning and evening rituals of feeding, the contented sounds of the hens as they ate under the anxious and authoritarian provision and supervision of the rooster a part of my daily rhythm, and helped to establish my subsequent association of those sounds with sheer serenity, as a signal that things were going as they should in the world.

Alice's mixed heritage endowed her with a coloring which I have never seen before or since. She grew into a beautiful hen, whose buff

(Orpington) shade was bronzed somewhat by her Rhode Island Red pigmentation, and whose neck feathers shone with some of the Oriental iridescence so loved by the Rhode Island Red fanciers. Her beak and legs were paler than the bright orange and yellow of the Rhode Island Red.

As though she were proudly conscious of her unique coloring and her double inheritance, rather than embarrassed by it (as Virginia probably would have been), Alice developed a confidence, a calm but commanding manner which got her what she wanted from her fellows. Without getting all in a huff, or throwing her weight around, she stepped into the flock of eating chickens, began to peck wherever she chose at the grain on the ground, and the others made way for her. Nor did she make them scatter in fright; they merely, without question, moved over. She was at the top of the pecking order, knew it, and so did everyone else; it was a fact of life and required no comment or reassertion. The fact that she was my choice and given certain privileges only reinforced Alice's preeminence, in contrast to Virginia whose position deteriorated partially because of her special treatment.

Her superiority was not confined to her own kind but carried over into her relations with me. Where some hens would flutter or squat, as Ruth, the Barred Plymouth Rock always did, or try to avoid being picked up like poor Minnie of the crooked tail, Alice never lost her dignity. She would not stoop to running away, and never indulged in an abject flattening to the ground. Nor did she come running when called as Virginia always had. She merely stood still, waited for me to pick her up, when, just as I took her, she would give the slightest indication of a dipping curtsey.

At our new house where we moved when I was six, Alice had more luxurious quarters: the hen house was set against the backyard fence, and the pens extended onto an adjacent vacant lot. The flock was augmented to a dozen or more hens of different breeds: a Black Minorca, Rhode Island Reds, Plymouth Rocks, White Rocks, a Leghorn or so, and a couple of mixtures. Alice ruled with seemingly effortless equanimity over the old as well as the new arrivals. If there were chicken grumbles over her authority I was never able to recognize them as such.

Cindy, our German Shepherd, was only about eight weeks old when Alice and she first met. Alice seemed to enjoy being carried about as much as Virginia had, so I frequently brought her into the backyard to play with her, to enjoy watching her find grubs and worms in my grandmother's flower beds, as well as to feed her extra grain from my hand. Alice, who had never encountered a dog before, quite typically showed no fear at all when Cindy bounced over with enthusiasm to greet her. She did not flutter, nor run away, nor utter a squawk. She merely stood her ground, fluffed her neck feathers twice their normal size and elevated her wings. As soon as Cindy came close enough, Alice calmly but with determination and great precision whopped the pup on the end of her black nose. That was all. Down went the feathers as Cindy let out a small yip and sat back momentarily on her haunches in surprise.

Alice's even temper disappeared only when she was setting, and during the following period when she was shepherding her youngsters through their babyhood. Then she was formidable even to me. Cross and irritable, quite ready to puff up to her most frightening proportions, she refused to allow me any privileges with her or hers. Like many compulsive, brooding hens, she would be so intent on what she considered the requirements of her setting responsibilities that she would refuse to get off the nest to eat and protested having to be lifted off so that her legs could uncramp. She would always hurry back as soon as she had eaten a token amount of grain, and with great care stepped back on her nest (one of the laying bins in the chicken house), turning over her eggs one by one, and then fluffing the warm insulation of her feathers over them all.

It is sad that the joys of knowing personalities like Alice are not now common for children. For in due time keeping chickens in one's backyard came to be frowned upon by an urban society which considered itself to be too "sophisticated" for such rural vestiges. We received one day a telephone call from an irate neighbor on the corner, a minister of the gospel, who objected to our rooster's crowing in the early morning. My father was forced to the conclusion that he had better remove the chickens to Rollinsville. So I lost my pets to the hills, and could enjoy them only when the family was able to go there for weekends during the school year, and for longer periods in the summer.

Alice and Cindy in their new living arrangements, found themselves frequently face to face without benefit of a fence between. Cindy was not a vengeful character and never seemed to carry a grudge. But it is to be assumed that she had not forgotten her original encounter with Alice; also that she was probably intrigued when she could not seem to budge Alice if she sometimes dashed into the middle of the flock, ostensibly on her way to something else.

When Cindy and Alice and the rest of the chickens moved to the hills they were left in the charge of an elderly woman whom Dad had hired that summer as cook for his ice crews. Mrs. Russell had nowhere to go after the icing season, and to their mutual benefit, she and my father agreed she would stay on at the Ranch/Ice Plant. She did, through that winter and for some years after, an astounding example of physical fortitude and stamina. Mrs. Russell loved all our animals, the wild flowers in summer, and did not even mind the outdoor plumbing, the twenty-below-zero weather, the coal stove, and the pump which provided the only water in the kitchen.

She adored Cindy and enjoyed the chickens (especially my Alice), whose twice daily feeding involved her crossing the creek and walking about a quarter mile to the chicken house. On one of my father's frequent trips to take groceries and to check on the numerous details necessary to keep the Ice Plant in order, Mrs. Russell reported with great sorrow that she had gone over to give the chickens their afternoon feeding, and had found Alice lying on the ground and Cindy standing over her. There were no signs of a scuffle, nor that Alice had had her neck broken or been injured in any way. We could never know for certain if Cindy actually had killed my hen; or if she had tried to play with her, and if because of her advanced age of some eight years, Alice might have had a heart attack. Cindy appeared, according to Mrs. Russell, merely to be looking at her in curiosity but with no appearance of contriteness. I must, knowing Cindy, believe the latter.

Alice possessed quite a fair share of brains, intelligence and personality as had Virginia, Ruth, and Minnie. I boil when someone says to me, "Oh, chickens! They are so stupid!" It is hard enough on domesticated geese and ducks and chickens, that they have been bred into anomalies who still have wings but cannot fly. But chickens have been the victims of the most ruthless, callous treatment of all. (One needs only to contemplate how our industrialized chicken factories traffic in living, breathing flesh.)

The core of society's collective tragedy today is that we have withdrawn ourselves so far from close and meaningful association with other forms of life, and accepted instead love affairs with gears and computers. Thus we have permitted the once rightly respected chicken to become the automated egg and meat machine to which it is now reduced.

Someday, perhaps, if there are still breeds of chickens cared for with decency and respect, cultivated for their beauty of plumage, their conformation, their rich brown eggs or even their big creamy white ones, admired for their individual characteristics such as fanciers praised at the turn of this largely unhappy century, others may be privileged, as I was, to grow in the joy of friendship with another intelligent and personable Alice or Virginia.

Chapter Four

CINDY

There have been three dogs in my life whose characters, intelligence, and love have in great measure shaped my thinking and my beliefs: a German Shepherd, an Irish Setter, and a Scotch Collie.

I was a mid-teenager when my father decided to purchase a blooded German Shepherd. He was at that time expanding his operations to include not only the natural ice business on which everything else depended financially, but to begin ranching on a small scale. Raising pedigreed German Shepherds was to be one area. It was an exciting day in early spring when the family journeyed south of Denver to a small farm, long since ploughed under and asphalted over for the cloned suburbia and shopping centers which now threaten a megalopolis along the entire east face of the Rocky Mountains. A customer of Dad's in the seed and vegetable business had imported a fine German Shepherd bitch, and we were to choose one of the females in her litter, for the unheard-of, fantastic price of $75.00, an extravagance my grandmother never recovered from.

The pups were eight weeks old, fat and woolly, tumbling over each other, their outsized feet squashing the fellow they clambered on. We chose one whose over-all coloration was an iron grey, but who had a black muzzle which looked as though she had been dipped in soot half way up her face. Among the grunting, yapping mass of furry bodies this one, we thought, had more personality and class. From her very first hours with us she justified our hopes by demonstrating the initiative and intelligence that brought her to magnificent adulthood.

I do not recall that she caused any trouble during even her first night away from her mother and brothers and sisters. An old coat of Dad's was laid on the floor of the cellar beside the coal furnace, providing a warmth which may have consoled her, plus the smell of one of the humans who carried her away. The cellar was not one that would have provided much comfort to anyone except a cave man. There were two cement-floored rooms in its half-excavated, subterranean ugliness. One was the furnace room with the coal bin at the end. The other room was a laundry. In the laundry, doomed to half gloom even on the brightest days, stood the old Thor washing machine.

Cindy

Cindy, left alone that first night, explored her quarters in that dark cellar and discovered the newspapers placed under the Thor to catch the oil that occasionally dripped from the motor. Since that space was just high enough for a pup to walk under, she decided to use it for her toilet. She never needed one single lesson—from us—in housebreaking. It was actually beneath her dignity to need it.

Because of her cinder black muzzle, my mother christened her Cinderella E-I, and her Certificate of Entry in the Stud-Book of the American Kennel Club so read, Shepherd Dog, No. 644604. Her pedigree papers said that she came from a long line of German dog nobility, whose family boasted names like Zora of Costilla, the Baroness von Hitzfeldt, Count Roland von Arapahoe, Stonehurst Fawn, etc. And here and there crept names of common, plebeian source such as Heidi, Frisia, Anni, and Yank. The history of her predecessors said that she carried in her veins champion, blue ribbon show dogs as well as hardy, muscular work dogs, perhaps descendants of the European "wolf-dogs" of whom Caesar spoke in his Gallic adventures. We felt quite set up about the whole adventure of owning a dog with pedigree ancestry, though I do not recall that this pride was our dominant reaction to Cindy for very long. She was too vital, too full of curiosity, affection, and high spirits to be loved for anything but herself. However, there was an aristocratic dignity in Cindy's character which could not have come from anything but a distinctive and superior blood line. The aristocracy of Nature asserts itself, regardless of, and on occasion in spite of, situations which may make it difficult for that excellence to surface.

Cindy's deep intelligence aided her in very quickly sensing and evaluating the various members of her new family, in adapting to their requirements, and in making her own desires and needs known. She frolicked in her clumsy puppy fashion over the backyard, tromping some of my grandmother's flowers in the process. She also met, during her first early weeks, my trim, golden-feathered hen, Alice, who rebuffed Cindy's advances with regal discipline. Ever after, Cindy treated her with due respect and kept her distance. She apparently decided that Alice had no sense of humor and never in the years that followed when they both lived at Rollinsville did she include her in the game of "pretend-chase" which she sometimes played with the other chickens.

It was soon apparent that Cindy was outgrowing the rather narrow confines of our small backyard. As summer approached and we happily prepared, as usual, to spend the vacation period at the Ice Plant in Rollinsville, Cindy of course went along. On this trek

Cindy found herself on first one lap and then another because there was no room on the floor among my father's freight load of crates and bushels and cans and kegs. She was still too young and inexperienced to ride outside on the fenced running-board. This accomplishment she learned in due time, draping her big body over the hood, and looking very like the figurehead at the prow of a boat, her hind feet braced on the running-board. Thus, in June of her first year she made her change of domicile, one that was to be permanent. She never returned to the city.

What exhilaration she must have felt on her first arrival at the Plant, the thrill of a new freedom of space, wonderful perfumes in the pure air, and her family with her, with whom to explore all the fascinating gulches, ravines, woods and meadows. Maybe we communicated to Cindy some of the special intoxication, a sense of joy unrestrained, which was to me the essence of our arrival for the summer, when a happy eternity of three whole months stretched luxuriously before me! It was an emotion whose intensity I have seldom recaptured in more mature years. Is this really what human maturity means? Loss of shadowless anticipation of shining and undiluted pleasure?

I do not believe Cindy ever lost her early enthusiasms. Judging from her later actions and emotions, I think the tremendous satisfaction and happiness with which we settled into Sunnybank (the family cabin Dad had just built) for the next three months must have transferred their decisive effect to Cindy.

The three months of her first summer, it seems to me in retrospect, provided that crucial developmental atmosphere wherein her most formative experiences occurred. She became attached to and convinced that this cabin was the real place, her true home. It was the spot she came back to for the most important events in her life, the one she watched over with special guardianship. Though she was forced to spend much more of her life with others than with us, there was never any doubting that to us, and to Sunnybank, she gave her most solid and enduring fidelity.

The separations which we endured were painful beyond measure. They meant that Cindy had to attach herself to, and "make do" with pseudo-masters, a string of caretakers with a variety of temperaments and capacities for understanding. But they, even the best ones, and the ones she most liked, remained just that—pseudo. No matter how contented she appeared to be with them, following them about, doing their bidding, when we appeared on the scene for a weekend or a day during the school year, she dropped them unceremoniously the moment we got out of the car. With unrestrained cries of joy, and much running about and leaping in the air, she attached herself to us until the sad moment of parting occurred once again.

If it were a long weekend, when we could stay Friday and Saturday nights, she seemed to be able to distinguish on both of those nights as we finished dinner, that our departure from the dining room meant that we were going to the cabin, but that on Sunday evening we were heading for the car to leave her, going she knew not where nor for how long. We surmised this because on Friday and Saturday nights she joyfully waited for us as we put our coats on, and accompanied us to the door with much tail wagging, and her eyes alight with pleasure. But our same routine on Sunday brought a drooping head, or she remained lying on the rug in the office, and refused to get up and see us to the door. It was never necessary to have to tell her that she could not go with us. There must have been something in our own actions or words which she picked up and correctly interpreted, telling her what was coming. This was true if we had come on a Saturday instead of Friday, and therefore had only one night at the cabin. So she could not have depended on counting nor on a rhythm of memory.

One can never know if the cruelty of separation from Cindy's loved ones was counterbalanced by being allowed to run and have her freedom unimpaired. She was truly too full of life to have been content in a city backyard, but perhaps she would gladly have traded freedom of movement for more constant companionship with those whom she so deeply loved and needed. From later experiences with the two dogs in my mature life, I would without hesitation (if I could return and live over the years with Cindy), have kept her with us no matter how large or small the backyard and managed her exercise as best I could.

I know now too, from agonizing experience on the human level, how infinitely precious are the swiftly passing hours of this life, and what folly it is to permit the pain of separation which can be avoided. There are no compensations.

In a lifetime of intimate association with many animals, I have never known an individual with such superb mental endowment as well as perfection of physical beauty as Cindy possessed. She arrived in this world equipped with the strengths of muscle and intelligence, dignity and bravery for which her working breed had been carefully developed in Europe. She was not debilitated by the subsequent perversion of this heritage, by inbreeding for "points" in showdom, regardless of health and stamina and intelligence, a tragedy which has happened to almost every breed as it has become popular in the

United States. This whole sad canine picture is an excellent example of our obsession with "first cost economics" in every field of human activity.

Cindy's muscular development was healthy and strong. Her shoulders were broad, the muscles between her front legs so prominent they rippled as she moved; to feel them was to sense the power that lay there. Her chest fur was a dark creamy color, crisscrossed with numerous cowlicks. Her legs and her huge feet were also cream color. Her feet were noticeable for other reasons: being so large they brought into the house whenever there was snow or mud, a goodly share of it, charting her course through the rooms. Being light in color they also showed every bit of dirt that collected on them.

I often wondered if Cindy shared with me the aesthetic pleasure of the natural perfume she applied to herself. She enjoyed rolling on the hillside or in the meadow, coming up with the most invigorating, clean, fresh aroma of wild sage and grasses in her fur. I loved to bury my nose in the fur around her neck, as I took her whole front quarters in a loving embrace. To this day a pinch of sage, or a smell of camomile tea brings back that dear animal.

The fur on Cindy's back and tail was a flecked, iron grey. Only on her nose, her muzzle, was she black. Her eyes were a fairly dark brown, not the deep chocolate of either my setter or collie, but full of amber glints. They made her eyes sparkle, though the lights were fully as much due to her superb spirits, her love of life, her unbounded energy. She was all in all the most magnificent single specimen of life I have ever known, her unusual mental endowment, her capacity for understanding, for loving, setting the seal of near perfection on and crowning the entire physical structure.

There was never a need for any Obedience School training for Cindy. It would have been like subjecting Mozart to the usual piano lessons ordinary young people have to take. She went so far beyond those training exercises, in her desire to cooperate, her almost prescient sense of what was being asked of her, that even to suggest the usual "tricks" seemed like an affront to her dignity. In fact, *she* seemed to regard them as such, and obeyed perfunctorily. She said "please" with no great enthusiasm, usually a low growl, as though a vigorous "woof" was indeed for poodles only and beneath her. She shook hands with even less alacrity, often waiting until she caught a tone of impatience in our voices before complying. Then she would proffer her foot, but turn her head away as though to show her disdain for such childish behavior.

At the time when Cindy was growing up the charm of Maurice Chevalier was abroad in the land, singing in his lilting, teasing fashion, "I kiss your hand, Madame, your dainty finger tips, and while I kiss your hand, Madame, I'm longing for your lips." I don't know what perversity tempted me to see in Cindy's big paws, as she lay with them hanging over the edge of the stoop at Sunnybank, any similarity to "dainty finger tips," but it was sometimes fun to tease her, if only because she was so understanding that she immediately recognized she was being teased. Her facial expression as she looked down on me, standing just even with her paws, was at first so condescending then so embarrassed, as I sang the ditty to her, that I could not resist the humor of it all. So I take the blame for Cindy's special aversion to shaking hands. She, perhaps more than any other animal I have known, reacted to the very suggestion of ridicule with injured or offended dignity. But all of them, of whatever species, have had strong convictions of their own personal dignity.

She liked to retrieve but that was because *she* devised a game out of it, and to hunt out the hidden object was a challenge which she enjoyed. In fact, she taught herself that "trick" by participating uninvited in a game of catch which we were having one time. Thereafter we had to have two balls, one for Cindy to keep her busy and one for us. Occasionally, even then, the correct ball got mixed up in the wrong game.

Part of the pleasure either in the game of retrieving or of carrying was putting to use a talent which had been developed in her working ancestors, who had carried messages or medicines or other necessities for their masters for generations. She proudly bore her burden, holding her head high, her eyes sparkling with both enjoyment and a sense of importance. Perhaps the two are faces of the same emotion, for I believe everyone enjoys at least a modest sense of importance, of performing a needed function. It is one of the most basic pyschological requirements for mental health, and it is certainly not limited to the human psyche. Other animals languish no less than the human when cast off because they are no longer needed, useful, or loved.

There may have been an element of teasing when Cindy refused to play those games, perform those "tricks" she considered immature, or to obey on command something she could see no purpose in, because she certainly did enjoy teasing. If we could tease her, why couldn't she tease us?

She once played a joke on us which could have been extremely inconvenient, had her idea worked. In order to provide security for the numerous buildings at Rollinsville, my father had used some old lathing strips to make a series of one-by-six-inch wooden labels. Carved into them were Sunnybank, Tool House, Garage, Dining

Room, Brown House, etc. At one end was a hole to take the thong by which the key was attached. This was the key we carried to and from the cabin whenever it was left empty.

Cindy on her own had observed the wooden piece hanging from our hands and offered, by gently taking it in her mouth, to carry it for us. She was always pleased when we passed it over to her and trotted ahead, tail wagging, her head held high, the bar between her beautiful, strong, white teeth and the long skeleton key itself bouncing back and forth. We commented on the fact that the key must hit her jaw uncomfortably, because she swung it wildly at times. One day as we approached the ice houses, at the east end of which there was a large pile of used sawdust, she ran ahead and started digging furiously, still firmly holding in her mouth the lathe and its dangling key. With her big paws she could accomplish quite a hole in a short time. The sawdust flew as she dug, and periodically she stepped back to pull the mound away from the hole with both her front and back feet. Then she dropped the key in the hole, and nosed her small mountain of sawdust back into place. Having restored the area to its original contour all in a matter of a minute or so, she turned and came back to us wagging her tail and looking very innocent. Had we not witnessed her performance we would never have been able to guess where she had either dropped the key or hidden it.

There was no hint in her bearing that she considered she had done anything wrong, but neither did she offer any assistance in locating the buried treasure. She kept wagging her tail and stood aside watching closely while we tried to find it, as though she were enjoying our difficulty. Much less adept than she, we scratched and dug, and even though we thought we had seen the exact spot where she had excavated it was amazing how hard it was to discover the key! We did at last do so, but ever afterward when Cindy decided she wanted to carry the key one of us was detailed to follow her closely and never lose sight of what she was doing.

She employed a clever ruse which also had to involve her sense of humor. Her delight was obvious in having devised an opportunity to do a double tease. It provided a delicious sense of breaking the rules under circumstances for which she could not exactly be chided.

Frequently when the family came down from Sunnybank for breakfast, crossing the valley and the creek to the dining room, the cook would already have thrown out the garbage from the day before, where the chickens could pick out of it what they wanted, after the horses had finished, that is. If things had progressed to the chickens' turn, Cindy would take in the situation as we came through the railroad gate by the ice house, several hundred feet away. She would trot a few feet ahead of us, then, with nose to the ground as though she had caught a fascinating and demanding scent, would lope off in the general direction of the chickens. She made a few deviations so that the route did not look too deliberate. But somehow that smell always led directly into the midst of the flock and without lifting her head, still sniffing assiduously, Cindy managed to scatter the flock with a screeching and flapping of wings. Then the scent would lead just where some of the chickens had fled—but never too obviously. Abruptly, that odor petered out and she could then go back and forth anxiously, all the time never lifting her head though her dancing eyes were following the havoc. Those who had escaped her first onslaught were caught in these vagaries, as she zig-zagged back and forth. Somehow the trail led here and there until none but Alice (so long as she lived) had been left undisturbed!

Before we had gotten too close, or she had carried things too far, she gave up, and with an air of false and wide-eyed innocence so thick it could have been plucked off her nose, she trotted back to us and sedately accompanied us across the bridge with never a glance at the scolding chickens whose existence she pretended magniloquently to ignore. Had not she been taught it was naughty to chase chickens?

Cindy's companionship was never passive. She participated in whatever we were doing, either by active cooperation (or hindrance), or by initiating an activity of her own which required our attention.

During and after haying time in August it was a favorite recreation of ours, after the evening meal and before we had hiked back up the hill to Sunnybank, to wander down in the meadow. The hay was down, and the evening cooled as soon as the sun had slipped behind the range; the perfumes, clean and sweet, of the newly cut timothy and the wild, high country meadow grasses filled our lungs. Cindy seemed to be as inspired by it all as we were. We used to refer to this hour as her "silly time," for she would shed all vestiges of mature respectability and cavort like any pup, jumping high in the air to snap at moths or butterflies, or rolling luxuriously on the fragrant ground. She would pick herself up from such a cleansing, shake off the bits of hay, and decide to run. Dashing madly away, she would turn abruptly, and run full speed back to us, putting on the brakes right at our feet. Most of this was silent fun as Cindy barked infrequently. She would pick up and toss into the air anything from a feather to a tiny scrap of mown hay, to a long willow branch, which might lie along the creek. Anything was grist to her play mill. Except for these walks and at this time of evening, she would have considered it beneath her dignity to behave with quite such a degree of abandon. But this was the magic hour when sheer exuberance of spirits and her deep joy in our being together could find no more satisfying outlet than in that random, uncontrolled outburst of energy.

How invigorating and contagious it was to watch her! She somehow lifted us along with her to a renewed appreciaton of how beautiful life was, and how very "right" everything seemed in these hours. I am forever grateful to that blythe spirit for the joy of those moments, when my heart leaped with hers.

Cindy lived through two winters in the company of one of the gruff natives and his wife, the same who bent our mare Emmy's disposition. What long hours of waiting and yearning must have stretched Cindy's life to a dull blank. I do not believe this couple was unkind, but certainly her depth of intelligence and her great heart were wasted in their company, and, of course, were lost forever to us. So when Mrs. Russell came to cook and remained on the place for some three or more years I know Cindy was much happier. They quickly developed an obvious understanding and appreciation of each other and a deep mutual love. One habit of Cindy's which she and Mrs. Russell were fully agreed upon greatly annoyed my father.

The kitchen cook stove was a mammoth thing of black cast iron, which sat plunk on the floor and had a flue going up the back worthy of a furnace. On its right was the enormous oven, which could and did take a couple of good-sized turkeys, or eight to ten loaves of bread. A hotel range it was, with a warming oven above the cooking surface, a most happy invention, lost along with the blessed warmth of radiation from the old coal range era. Little Mrs. Russell, a thin, wiry 65-year-old veteran of mining and lumber camp outfits, was complete mistress of the intricacies and temperament of that stove which could not claim the niceties of thermometer or timer or regulation by any mechanics but Mrs. Russell's own expert judgment.

During the winter harvest, when Mrs. Russell was feeding hearty meals to thirty or so men, three times a day, and when the weather was 20 to 30 degrees below zero, the big stove was going full tilt, full time, and she was scurrying around the kitchen pulling loaves of freshly baked bread from the oven, basting a huge pork roast, mashing a small washtub of potatoes, etc.

When Cindy came in from the lake, having been following my father around, she would come immediately for a greeting from Mrs. Russell, and then throw herself down with her usual thump, sprawling before the warmth of the stove. This meant that she covered a goodly area in front of the stove, and her legs protruded another couple of feet. Mrs. Russell never asked her to move, but stepped over and between her legs, straddling her, to reach the top of the stove. Cindy accepted this indulgence without question, and only when Dad would scold her would she look up surprised. Then because he would insist on her moving out of Mrs. Russell's way, she would look at him with her ears back, hurt and embarrassed. Finally, with obvious reluctance and some impatience, she would pull herself up with exaggerated effort and slowly, very slowly, stalk in offended dignity, to the rug in the office, there to throw herself down with an audible groan.

Mrs. Russell would say reprovingly, "I don't mind stepping over her, Mr. Espy." She truly didn't, and Cindy knew it. So they continued. And so did my father.

Cindy knew other ways, too, to disagree with my father about what went on at the dining room which she had had to accept as home when the family were gone. She showed an intimate knowledge of our eating habits and those of the crews. According to strict orders from my father, Cindy was to remain in the kitchen with Mrs. Russell during mealtimes in the big dining room. She first succeeded in fudging a little by lying just inside the dining room at the end nearest the kitchen and farthest from our table. As was the habit in our crew dining room, the tables were set before the men entered. Desserts were already in place along with each man's silverware, plate, coffee cup, and water glass, and thus, after passing the serving dishes to those at their table the meal could be eaten at whatever speed a man wished. The speed at which the men did eat was about equal. One or two, of course, always left the table and the dining room in less than fifteen minutes, having bolted the meal, it seemed, in one gulp. The majority departed in a large group a few minutes later.

It was at this precise moment that Cindy timed her move. Under cover of the general exodus, and the noise the men made by pushing the benches away from the tables, she threaded her way among them, reaching our table unnoticed. There she quietly ducked and walked under it. A moment or so later, either my mother or I would feel a weight on our laps. There would be Cindy's head, pressing heavily, a pair of half mischievous, half pleading eyes looking up at us.

Her choice of signal was the important decision. She timed her move at what she considered the strategic moment of least visibility. I am sure Dad was aware of her presence under the table, and also that he let a few transgressions get by without calling attention to them, simply because he also enjoyed this evidence of her insight and her initiative. Since she never became one of those objectionable beggars from the dinner table (it would have been beneath her dignity to do so), he permitted those intrusions.

Breaking the rules by lying in front of the kitchen stove when Mrs. Russell was so busy was a joint effort by Cindy and Mrs. Russell; entering the forbidden dining room at meal time was Cindy's own refraction. But she considered that for our part we also broke rules of her rightful behavior. Because the big stove in the dining room kitchen radiated so much heat, it sometimes seemed as though Cindy did not mind cooking her brains by lying so close to it. But that was heat controlled, inside a container, and she saw or felt only the warmth. Far differently did she feel about an open fire, either out of doors or in front of the fireplace at the cabin. The sight of leaping flames brought out all the wild inheritance in her genes, usually so dormant in the Cindy we knew. When Dad built up a nice roaring fire of a chilly evening, she would slink to the farthest corner of the living room, back of a piece of protecting furniture, her eyes full of fright, her tail almost between her legs. Nor could she be enticed out except under duress. If we succeeded in getting her to the side of the fireplace, she would make a leap to the couch where we were sitting in front of the fire and crawl behind us, utterly miserable. As she was so big this usually resulted in our being pushed off. But as soon as she found herself with no barrier between herself and the fire, she dove off the other side and vanished to a bedroom. No amount of coaxing would bring her forth again. I am sure she never understood how we could try to break down her deep fear which forbade coming within range of open flames.

Such a fascinating bundle of combined responses, decisions and initiatives, comprehension, of love, of domesticated or man-induced traits, and a goodly share of her own wild instincts as Cindy was! Sometimes as I watched her dark eyes lit with those amber lights reflecting her eager intelligence, it seemed to me there was a spirit within and behind them that was much greater than the limiting form which encased her, as though there were something much, much more shining through, straining to escape.

One summer afternoon she had lain behind my mother's chair in the sunroom at the cabin, dozing serenely while we read and sewed. Tiring of our sedentary occupation, we decided to go down to the Plant about four o'clock, watch the icing of the refrigerator cars, and walk to the dining room with Dad when the shift ended. Happy to have the confinement ended, Cindy trotted close to us down the hill. As we got to the road, a piercing scream like that of a woman in panic almost filled the canyon with its volume. We all stopped short in our tracks, including Cindy. She apparently knew a lot more than we. From the placid companion who had been within reach of our hands a moment before, she became an untamed creature, unrecognizable and showing no cognizance of us. Her eyes spoke to an immediacy which demanded, as in ages past, retaliation or refuge. Her ears were pulled so flat to her head they changed, as do a horse's flattened ears, the contour of her face. She half crouched as though to spring, and throwing back her head emitted a wild howl so unrestrained, weird, that it made our blood run cold. I have never been privileged to hear the call of the wolf in the far north, but I am sure Cindy's cry was a close relative.

The first scream, it seems, had been that of a mountain lion—why at that particular time, or from where, we never knew. But Cindy had left us that instant and spanned two thousand years or more to reach into her past. It was all over in the space of a minute, the scream and her answer, but the memory will go with me always, because it linked Cindy so dramatically, so organically with her ancient lineage. It further deepened the integrity of her person, her right of place in the evolutionary scheme of things.

Such evidences of the labyrinthine paths which wind their way from eons past through every structure to its present state, cause us to ponder our heritage. These ancient lines were thus exposed in Cindy's rare reversions, although predominantly she was as domesticated, as "civilized," as we.

Though family trips to Rollinsville were infrequent and, of necessity, short in duration through the winters because Sunnybank was only a summer cabin, we did make the most of our brief hours with Cindy as she most certainly did with us. While Dad was living up there during the two weeks in winter which it usually took to complete harvesting the ice, she would accompany him faithfully all day, even trying to walk along the swaying, icy galleries in front of the ice houses when he went to check the storing of the layers inside. Cold and wind and snow notwithstanding, she was at his heels. Many dogs will not venture onto slick ice, but she learned to keep her footing, perhaps by her own sharp observation. She shuffled, just as Dad did, in very un-dog-like fashion, and managed to slip and fall very seldom. She became, in fact, rather sure of herself on ice.

She also managed, without previous experience or training or encouragement, to climb a full story up a rough ladder. She had followed us to a building being constructed in Rollinsville, and did not intend to be left alone. When we climbed the ladder to reach the second floor, we had expected her to wait for us to climb back down.

But we had no sooner started to walk over the sub-flooring of the second story than we found Cindy beside us, more than pleased with herself. Because there was no way to get down except by the same method by which we got up, Dad had to pick her up and carry her down, no small problem considering her size and weight. But how she had negotiated the rungs of that ladder without falling through we will never know. What a shame we missed it!

Fortunately, almost all of our activities were within easy reach of Cindy, and some of them provided her with as much unalloyed pleasure as they did us. I recall as one of the string of idyllic episodes in my youth, the horseback rides we took in that mountain country, on which Cindy always went along. During that time, the Colorado Rockies were by and large at an in-between stage of human occupation. The hectic mining activity which characterized the last half of the 19th Century had largely dwindled away into history. This was due to the final realization that the "boom" bred from the California find of 1849 was more "bust" than anything else. The human ants who, from the 1859 discovery of gold in Colorado had swarmed up and through every canyon, every gulch, had retreated in large part to the cities and the farms on the plains, to make a more reliable, if less exciting, living. The mines were long since abandoned, and the mills, the cabins, the roads, entire towns had given up to the will and the workings of the seasons. Hence, when I was growing up, there were open, unfenced mine shafts to fall into, cabins whose roofs sagged and whose doors stood forever half open, revealing old newspapers, rusty bedsprings, and the musty evidence of generations of mice and chipmunks. The rather meagre roads were still bare of vegetation after half a century of disuse, and they made excellent horse trails, if serving no other purpose.

I was too young and ignorant to observe and interpret what these scenes really portended in the American character. I only know I loved the deserted towns, and found beautiful the trails winding up and down gulches and around the once mine-denuded hills now bringing forth a second, smaller growth of forest. At the hot noon hour the air would be pungent with pine perfume. As the forests grew again, the birds had returned and established their pattern of living. They sang through the trees during the mating and nesting season, the Swainson's and Hermit thrushes trilling their electrifying arias with such exquisite clarity that it made goose bumps stand out on me. Bulldozers and subdivisions, like a measles outbreak all over the East Range had not yet created the second "rush to the Rockies." A few people maintained modest summer-home cabins. One could ride for hours and miles, without encountering a fence, a person, and certainly no ORV's tearing up the fragile soil in roaring disdain for the environment. An occasional herd of cattle grazed quietly. Around our part of the mountains, a good part of the public lands became National Forest early in the century. And the Forest Service was not yet, at that time, mainly in the timber business.

So I suppose one could say we had the best of two worlds: we had more or less harmless, primitive wagon roads which made riding comfortable, with as close a return to a condition of nature as could be achieved. One of our favorite horseback rides each year was to Corona, on the top of the Continental Divide, a round trip of some 25 to 30 miles.

There was probably no way for Cindy to know whether the expedition was going to be short or a day-long one. Since the ride to Corona was made only once or so every summer, toward the end of the season, there were no clues to which she could pin her judgment. Hence, her enthusiasms and her energies, which were always enormous, and spent with glorious abandon, were expended as usual as we went along. She would make her first dash into the woods as soon as we left the main road, to emerge out of nowhere alongside us a half-hour later, breathless, her tongue hanging out so far I sometimes thought she would lose it. As soon as we came to the faintest hint of moisture in a shallow mud puddle she would throw herself on it, then get up, plastered like a rhinoceros after his mudbath. Her lovely buff-colored underside, her legs and her feet would look hopelessly filthy and caked. But a metamorphosis always happened as soon as we came to a stream, however meager. She would wade into it joyfully, lapping noisily, splashing around as though on purpose to bathe. If there seemed to be time, she would lie down, sprawled at full length in the stream, letting the water eddy around her, soaking up as much as she could in her thick fur. She would get up from this spit-bath amazingly cleansed, ready to repeat the whole cycle. On the final home stretch she would trot wearily alongside the horses, occasionally running a few yards ahead to throw herself down, snatching a few seconds of rest until we caught up with her.

For a few days following she limped around because her feet hurt, the pads having been worn absolutely smooth and somtimes even cracked. But her healthy constitution repaired such damages in a minimum of time, and she was soon back to her buoyant self, ready for play, hide and seek, a hike, or even another adventure with us and the horses.

One of our most beautiful rides took us on an old mine road, through a gulch which offered a stream choked by willows and alders, delicious cool spots of moist black earth, groves of fir and spruce where there were for Cindy enticing scents of numerous animals. On the drier northern slopes were open stands of mature ponderosa pine. There was also an abundance of wildflowers. If we rode this trail in mid-summer, there were magenta shooting stars and blue and pink giant mertensia along the stream, mariposa lilies in all their simple purity on the open slopes, and the long stemmed, lavender-blue and white, exotic Colorado columbine regally straight against the white bark of aspen trees. These choice flowers luxuriated in the aspen groves where the rich mulch of generations of fallen leaves gave them prime soil in which they could grow, tremulous and thick, and enjoy at the same time the delicate shade they needed from above. Many times Cindy would stop to smell a wildflower gently, and I often wondered how much aesthetic pleasure she derived from it. How greatly intensified must have been the perfume to her sensitive nostrils beyond what I could ever know.

Cindy's carefree life, her unchallenged position as the only dog on the place was abruptly changed in her third summer by the event for which my father had originally purchased her. Through the same source from which we had acquired her, Dad located what he considered a suitable male, and Cindy was on her way to becoming a mother.

This phase of Cindy's life was one which merited the observations of a psychologist. Or, on reflection, maybe it didn't. I fear a psychologist would have divested her behavior of all significance and reduced it to sheer instinct, or reflex, or conditioned response. All of which it plainly was *not* to us, nor to any of the numerous people who knew and lived with Cindy. Science, no less than religion, but under differing guise, seems constrained to establish the absolute supremacy and uniqueness of man. It also feels at its present stage of development it must be able to quantify all data. If science is unable to assess the unassessable, then, ipso facto, the unassessable cannot be so.

Cindy's first litter numbered six, and they were born just before we arrived at Rollinsville for our summer stay. There is probably no more joyous—nor instructive—experience for children than the participation in a litter of newborn babies, whether they be kittens, gerbils, rabbits, chickens, calves or whatever.

We did not know immediately that Cindy was a mother, only that her greeting was strangely brief. Whereupon, she disappeared for a couple of hours. When she reappeared, the reason was clear. Her teats were swollen, and the cause of her preoccupied air was confirmed. But where did she go? Following her did no good because she was obeying an instinct deeper than her domestication; she did not go to her lair, but waited until she had either given us the slip or we were otherwise engaged. Finally, as we sat before the fire at Sunnybank an evening or so later, Dad thought he heard a whine from directly underneath us, and so the hunt narrowed.

The cabin was built on a 30 to 40 degree slope of the hill, and at that time Dad had not yet been able to afford the rock foundation he planned. The house stood on a number of stout 8-by-8 timbers set in concrete bases. The entire area was open underneath. The big moss fireplace formed the core around which the house had been erected, and it rose from the ground itself. Because of the slant, there was still an accessible area, however slender, behind the base of the fireplace.

Dad decided that Cindy must have belly-crawled into this tiny space, to have and keep her pups. So early the next morning when Cindy made her first appearance for the day, we trooped below, Cindy trotting alongside us. She watched with the most intense interest. Considering that she had stolen her hide-away, she surprisingly showed no apparent resentment as my brother, then a slight youngster of 10, crawled up behind the fireplace to see if Dad's surmise was correct. Cindy made no move to obstruct and acted indeed as though she were as much one of the spectators as we were. The pups, as so many wild young behave in the presence of danger, made not one sound.

From the depths of the darkness my brother announced in a muffled voice that there really were some pups, and he began to hand back each puppy singly to me, half-way along the route. Dad put them gently on the ground, and as each one emerged, crawling blindly and unsteadily around, Cindy inspected them, nudging them over on their backs, with an authoritative and possessive lick. There were five thus far, grey and black butterballs, with short, thick but very wobbly legs. Then my brother said in a different tone, "Oh, oh, here's a cold one."

"This is what I wanted to be sure of," said Dad. "Hand it out."

The tiny, still, cold body was laid on the ground near the wriggling mass of warm and now whining babies. Cindy moved to inspect it. She sniffed it carefully, rolled it over, then without hesitation took the body in her mouth and turned away. As we stood silently watching, she walked out from under the house, her dead baby dangling, and we followed to see where she was going and what she was going to do. She almost seemed to have her destination in mind, as she went directly down the hill a short distance, to the base of a large ponderosa, and there, with the puppy still in her mouth, began to dig a hole. It was rocky and the going was not easy, but she soon had a little hollow, and into it she gently laid the body, then began pushing the dirt back into place with her nose. After a final scrape of the sandy soil, she gave a prolonged sniff.

Finished, she turned away, glanced at us lined up as at a human funeral with our eyes brimming with tears, and returned in quiet dignity to the remaining litter, now wailing loudly for attention. There she inspected her living charges again—and life went on.

My father's decision to put Cindy and her brood in a small dog house down near the barn may have been a major factor in the attitudes she later displayed, but of this, of course, we could never be sure. This much I do know: she was quite capable of connecting the fact of her progeny with the isolation from Sunnybank and her family. As extremely sensitive to mood, tone of voice, facial expressions as Cindy was, and with her large vocabulary of comprehension she could not fail to be acutely conscious of our absorption with the puppies. She was, moreover, very possessive of her human family, and had never before had to share our attention with another dog.

At any rate, I can report the facts as they happened. Dad, very wisely, would not allow us to handle or play with the pups for a week or so, until they opened their eyes and were more firmly grounded in their hold on life. During that time, Cindy would spend an hour or so with us, then disappear for a while, returning with question marks in her eyes. "What have you been doing, and where have you gone while I had to be away? What have I missed?"

The magnet that draws people to small puppies, or the young of all species to each other, was very potent in our case, and as soon as we were allowed, we visited the dog house and began to play with and fondle the wriggling, black and grey fluff balls. Cindy's reaction was not at first recognized inasmuch as it was unexpected, and I guess we were very dense. She stood aside and watched our preoccupation with her progeny. They were not in her mind, it developed, an extension of herself, or a part of her on whom we were vicariously lavishing such affection. They were separate entities, and when we were preoccupied with them, she was being ignored.

She began that first day of our attention to her pups, to attach herself constantly once more to us. She did not absent herself as she had been doing, to nurse and care for her pups. We could not help but notice this, since we had not been near the dog house during the day and neither had she. By evening we all set out from Sunnybank for the dog house, Cindy accompanying us happily. From some distance away we coud hear the plaintive cries of hungry babies, but when we got there, Cindy refused to go into the house. She showed no interest whatsoever in their distress, their empty stomachs. Finally Dad pushed her into the house, which resulted in Cindy's snapping at the pups as they whimperingly sought her full nipples. She refused to lie down, but dropped the pups unceremoniously, strewing them yapping in distress over the ground as she deliberately came out of the house and walked away.

We were grossly unfair to Cindy because we failed to comprehend why she felt as she did. Dad grabbed her and led her back, crossly forcing her to lie down and nurse the pups. She obeyed, with the most obvious reluctance, an expression of resigned martyrdom and mute accusation in her expressive face. This unhappy performance had to be repeated at regular intervals. She gave the puppies only the most rudimentary, infrequent inspection and clean up. She would lie there holding her head away from them in disdain, until we considered they had finished nursing. Then, abruptly, she would get up, sprawling the pups wherever they fell, and stalk off in haughty dignity. Finis, until she was forced to repeat her duty.

Over the summer as the pups grew and were able to follow us, to lope around, to try to trundle after their mother, we watched this relationship of reluctant motherhood and rejection of offspring intensify. A well-defined pattern of reactions set in. Dad had decided early supplemental feeding was inevitable under the circumstances, so a portion of the morning milking was set aside for the pups each day, and Cindy took advantage of this to wean the youngsters in record time. From that time on, she demonstrated unrelieved boredom toward them, which progressed to dislike and then to outright rejection, whenever one of them approached her.

When she came upon them she would growl menacingly, and advance upon them as though to bite. In fact she did so until they learned to avoid this unnerving experience by cowering, by rolling over abjectly onto their backs, feet in the air, whereupon she would disdainfully turn her back on them. But they did not dare to rise until she had left the area because if they did, her discipline was repeated with more severity.

One might say that Cindy became a distorted personality, a dowager with the most intransigent, dictatorial motives. To the extent that jealousy prompted her action, this might be true. But predominantly it was simply a desire to be rid of, or to get out of the way of, the offending presence and the noisy demands her offspring made. That accomplished, she had no further interest or desire to control them. They could go their own way; she was supremely indifferent. But in her presence there was to be no question that she alone ruled supreme. For in her mind, her supremacy successfully reinstated her claim to our sole love and attention, and negated the existential fact of their connection to her, their right to make demands upon her. She had thus forthrightly and honestly dealt with a situation which brought her pain. Moreover, it is quite possible that she did indeed have a lesser degree of maternal feelings than many females. If this is so in the human animal why not in others? The variation in so-called maternal instinct is in fact a phenomenon well-known to stockmen. Cindy's problem was complicated and intensified by her love for us.

We were culpable in not recognizing early on the psychological depths Cindy possessed, and thus avoiding a chain of reaction that, once having set in could not be reversed. But she, even in her pain, never turned her resentment against us, even when we held or fondled her puppies. She longed only to reinstate the old, mutually exclusive bond of love and companionship which was so plainly the guiding light of her life.

And what of the pups who were the center of the controversy? What effects on their development and their characters did Cindy's rejection and intimidation of them have? So long as any of the puppies in that litter, or in the other two which she had during her lifetime, remained in the area, there was never a moment's doubt or misunderstanding about Cindy's law, that her young must do abject obeisance whenever and howsoever long they were in the presence of their mother.

Whether she would have been a timid and ingratiating individual anyway, we cannot know, but Biscuit (one of Cindy's first born whom my father gave to Andy, the foreman of the Ice Plant), remained to her dying day a rather apologetic personality. When Andy occasionally brought her with him to work and she encountered Cindy, long after she was full-grown, she would cower upon seeing her mother, then finish by lying down and rolling over on her back, as Cindy haughtily inspected her and allowed her grudgingly to pass muster. If Biscuit tried to rise too soon she was put in her place by a quick growl and a lunge.

Cindy's last litter, two or three years later, numbered only two, whom we named Hans and Gretchen. (After that Dad gave up on the idea of establishing a line of German Shepherds. Cindy was no loving and protective matriarch.) Hans was a happy, well-adjusted, friendly reproduction of his mother in color and conformation. Gretchen, however was, we always believed, her father's child. For Cindy had broken out of her hated confinement during heat, and had been seen in the company of a coyote. Gretchen's coat, a light brownish grey mottle, was that of a coyote, her eyes the same light brown, and her whole behavior that of a captive wild thing. Her mother's domination did nothing to alter her outlook, and she was of course abject in her presence, either ignominiously lying on her back, or crawling on her belly to make obeisance. Her gait was a slink or a lope. Instead of trotting as Hans did, tail and head up, when she reluctantly was lured into the house with Hans, she quickly darted into the darkest room in Sunnybank, diving under the big double bunk, clear to the farthest corner, there to remain in spite of coaxing. Meanwhile Hans would throw himself on a rug in the middle of the room, and aside from carefully avoiding his mother's displeasure would cordially accept any family attentions, his big, powerful tail thumping the floor noisily.

The little porch entry to Sunnybank became from her very first summer, Cindy's post, her spot, where she could survey the entire canyon, the operation of the Ice Plant, see anyone coming or going near the cabin, and, most important, be immediately on hand for any activity on the part of the family; or in the way, so that no activity could go on without her. So much did this end of the stoop become Cindy's that one almost saw her there when she was actually somewhere else. It was like the old wicker rocking chair my grandmother sat in, in the cabin living room. No one else ever sat in that chair; it seemed always to be occupied by her whether she was actually rocking in it or not.

That stoop was for looking, watching, for napping, and for waiting, yes, for waiting. The long days that followed the family's fall return to Denver were the most pitiful and painful proof of Cindy's desolate waiting, and revealed how she dealt with the pain of separation. Mrs. Russell during her years at the Ice Plant, told us that Cindy would leave the dining room as soon as she let her out early in the morning, trot over the bridge and cut across the upper meadow, through the fence, over the railroad track and the highway, and up the hill to Sunnybank. There Mrs. Russell could see her climb onto the little front stoop outside the kitchen, throw herself down at the far end, her big paws hanging over the edge. And there she waited—returning to the dining room and Mrs. Russell only at night, to eat and sleep, and resuming her waiting and watching once more the next morning. When she had finally convinced herself, after a couple of weeks or so, that this was the end of the summer companionship, she would return to the dining room to spend her hours in the company of those who fed and housed her that winter.

It was Mrs. Russell who most genuinely sensed the pathos of Cindy's patience, her gradual resignation, the lowered tempo of her enthusiasms.

Mrs. Russell took a well-earned retirement in 1932, and a niece and her husband, a former Western Slope rancher, came to the Ice Plant to be cook and caretaker. Mr. and Mrs. V. assumed the care of Cindy with almost as much sensitivity as Mrs. Russell had shown.

It was in the fall of 1933, when Cindy was seven, that the first and last illness in her life occurred. It was assumed at the time, or at least so my father had been told by the people who sold Cindy to us, that pre-natal inoculation of the mother for distemper assured immunity for the pups. That this was not and is not so is now generally recognized. And it was apparently distemper that Cindy had contracted. The disease moved rapidly.

When we came to the Plant for a Sunday early in November Cindy did not come bounding out to meet the car. Instead, we found her inside, in her bed-corner of the kitchen under the high drainboard beside the big sink. Drained of all enthusiasm, she barely dragged herself out to greet us. After saying hello she turned immediately back to her bed, her eyes heavy, her nose hot and dry. She simply did not have the energy or the will to follow us about, and only thumped her tail when we reached in to pet her or to talk with her.

As we got ready to leave for Denver, our hearts heavy with worry over her, she came out of her little shelter, sat down, and with the utmost gravity and dignity, offered her paw. As I have explained, Cindy had always regarded the act of shaking hands as beneath her dignity, a feeling reinforced by my teasing her. She had therefore, never in her life initiated an offer to shake hands with us. But she was this time almost ceremoniously and significantly doing so. Two days later, Cindy passed away, November 11, 1933. She was buried by Mr. V., at the foot of a young ponderosa pine, close to Sunnybank and just above the big pine where she had buried her puppy years before. He lovingly placed a big stone at the head, and outlined the grave with smaller rocks.

The relationship we knew with this magnificent fellow creature, was in its fullest sense a partnership, a love between equals. She shared everything she had to offer and absorbed all the love we had to give. But in her total and whole-hearted giving there was never a slavish abjectness in her manner, she simply gave without stint out of the richness of her nature, her intelligence, and her heart, and accepted our reciprocity with joy. She revealed to me dimensions which increased my awe and adoration of the uncountable significances in the web of life.

Cindy's story does not quite end with the little grave just below Sunnybank's walls. And the sequel, I think, is quite in keeping with the intangibles of that web, about whose workings we know so little.

A year after Cindy had so knowingly said "goodbye," my brother and I bought a Scotch Collie pup for my mother for Christmas. Tammy grew to be a very handsome dog, though never robust in health and was, moreover, a "city squirt," quite removed from the rigors and necessities of mountain life. He was perhaps a year old when we made one of our weekend trips to the Plant where he had come many times before.

We had finished dinner at the dining room and walked leisurely in a lovely moonlit night the quarter mile across the canyon and up the gulch to Sunnybank. Tammy was trotting ahead on the winding road, aware of our destination. As we approached the cabin, Tammy was about ten or fifteen feet ahead, and ready to put foot on the stoop when he stopped short, a low growl coming from deep in his throat. His ears flattened to his head, he spraddled his legs, his fur stood up along his back, and he backed up to block our coming onto the porch. He was staring fixedly at the far end of the stoop. As it was bright moonlight, we could see also. There was absolutely nothing—*nothing*—on that stoop. After a few moments, Tammy's obsession seemed to loosen its hold, his fur lay back down, and he turned to look at us like someone being roused from a dream.

We all took a long breath, as though released from a spell, and went into the cabin, silently shutting the door. But I have never been able to shut the door against the conviction that on that spot where she had lain in the summers we spent together, on the spot where she had kept lonely watch for the weeks after we left, on that spot where so much of her great heart and soul had spent itself, there the essence of Cindy had been lying that night, with her big paws over the edge. And perhaps before that, and perhaps many times since. I only wish I could have seen with the clarity Tammy did that night.

Love like hers, comprehension, understanding that passed beyond the capacity of words was and is too real ever to vanish into nothingness, and I shall meet what Cindy was and is again, someday, somehow.

Will she be lying on a similar outlook, waiting for me? Or will she be sitting on her haunches, with her paw ceremoniously and willingly held out to me in recognition of our reunion?

Emmy

Chapter Five

EMMY

Emmy—the name did no justice to the grace and comeliness she possessed, but was given to her in a streak of perversity by Andy who was caretaker at the Ice Plant the year she was born. It did, however, seem later to be appropriate to other portions of her character—stubborn, wayward, self-possessed, mischievous.

She was the first colt born at the Ranch/Ice Plant, and she enjoyed the attentions usually lavished on the first-born. Some of them might happily have been eliminated, like the teasing and provocations which Andy was guilty of. His actions when she was a very young filly had a lasting effect on her disposition, marring an otherwise serene babyhood, and ensuring that she would regard almost anyone, human and otherwise, with a measure of distrust which manifested itself in a perversity of her own, and a calculating intent to reciprocate early torments.

Among her own kind, and throughout her life, Emmy enjoyed a privileged position. Pampered by a very protective mother, through her early years she enjoyed also the interest and indulgence of the older draft horses. After they had passed away, she became the eldest in rank of the following additions and employed with assurance all the prerogatives of seniority.

Patsy, Emmy's mother, was a pretty sorrel riding mare, in foal when purchased for my mother. Gentle, obedient, undemanding—her wilful and capricious offspring must have tried her patience many times. As an over-indulgent parent, Patsy allowed Emmy to suckle until she was so large she practically had to kneel to get to her mother's nipples. It became a security routine, perhaps in its way an equivalent to sucking one's thumb.

Emmy's coloring was as plain and unremarkable as her disposition was otherwise, a dark, even, chestnut brown, with mane and tail of the same soft hue. She was close-coupled, small and dainty-hoofed, a springy walker, carrying her head always erect, arching her deep, full neck. She never walked with her ears anything but up front, and her large eyes were quick to spot any possible fun or mischief. She was largely Hambletonian, with a strain of long-ago Arabian background via the Spanish and Indian mustang in her small, delicate muzzle, the forward set of her dark eyes, and in her soft ears, with their tips pointed inward. Her stamina and her canny intelligence, as well as her high spirits, could well have been part of her Arabian inheritance, while her spunk and contrariness were more the mustang addition of later generations. She invariably returned from a long and arduous trip with energy to spare, her ears pricked, head up, and her step high, when the other horses would be stumbling, their heads down with exhaustion. Indeed, most of the horses would return from a particularly rough day over the rocks and forbidding terrain, with one or more shoes missing or loose, because of slipping on stones or stumbling. Emmy lost her shoes only by wearing them so thin they would crack or fall off because the nail holes had simply worn too big. She was a comely piece of horseflesh by any man's standards.

From the typical, gangly-legged colt, maturity brought her the graceful and rather delicate legs which were so pronounced a feature of her appearance. For some years after her emergence from inquisitive and uninhibited mischief-making as a young filly, Emmy put on weight so that she was, during most of her mature years and into old age, more of a butterball than not. Since she was so close-coupled the roundness was even more apparent, and certainly more noticeable to anyone trying to saddle or ride her. For, as would be expected, she learned early to take a deep breath and hold it while we cinched her bellyband, resulting for her in a pleasantly loose saddle when she thought we were through. We dared not try mounting her without a second cinching, timed with an accurate appraisal of her inattention to the matter. But since her backbone was lost in the round fat of her back, her saddle never stayed where it should. If one climbed, it seemed to slide perilously toward her rump. If one rode downhill it was, before long, like riding over her front legs if not her neck. There were always stops on a long ride to cinch Emmy, long before it was necessary to check Kim or Ginger, or Cassie or Booker. I feel sure Em had her own private means of loosening that saddle which we were never able to detect, even after the first tussle.

Emmy's native endowment of mischief, the good-natured indulgence of the old draft horses, Chub and Tuck and Floss and Tom, who allowed her to nip and chew, the long months of being her own boss during the winter when no one deterred her from following her curiosity, wherever it led, all ingrained her independence and her wayward tendencies. She was an equine "Fille du Regiment" in a manner of speaking, and just as irrepressible and irresistible. Her adventures as a filly with the draft horse pensioners and her mother up on the Continental Divide the summer Chub led his string up there had further entrenched both her sense of freedom and adventure, and her privileged position in the herd. This made for a rounding of her disposition, but not exactly for a docile, non-assertive character desirable in a riding mare.

The high country experience no doubt added to her native hardihood, but also to the exercise of her common sense, of which she had a generous supply. In all her 28 years I never knew Emmy to be really afraid of anything. Her sense of play and mischief sometimes made her appear skittish or apprehensive or momentarily panicky but this was either because she was in high spirits and such antics worked off a certain amount of excess energy, or she knew she was teasing one of the family, or even better, frightening some stranger.

Early in her career she observed and learned from Chubby how to use her mobile upper lip, and her whole delicate lower muzzle, to open the screen door at the dining room, and this was a means thereafter of gaining entry to numerous places and things.

To Emmy, a door became what a door should be. If she could open one or go through another, all doors should likewise be operable. Hence she never hesitated during rides my brother and I used to take, to digress from an old path we were following, and like an explorer, walk unhesitatingly up to a miner's deserted cabin and either push the door with her nose, or pull at the edge with her mobile lip. If the door were missing or merely sagging she would poke her head inside and look around in concentrated curiosity, her Arabian ears with their in-pointed tips almost touching each other. Rotted wooden floor planks did not deter her. And if I had not reined her in she would no doubt have stepped clear inside. Scraping me off in the process would not have mattered, or maybe she had figured in that possibility as a side benefit!

My brother had a penchant for peering down dangerous, forsaken mine shafts left open for half a century. He also liked to venture into abandoned mine tunnels which were equally as perilous because of their imminent danger of caving in, and his mare Cassie seemed not loath to join him on both investigations. These relics of the first gold rush days Emmy very wisely considered out of her range of interest, as she had a healthy regard for her own safety. And she absolutely refused to approach them, not hesitating to break a rein if urged beyond her instincts.

However, she also enjoyed certain apparent lapses of her eminent common sense, as a prerogative of feminine unpredictability. She could be the most flighty of any of our horses if she spotted a scrap of paper being blown along the road, or spied a culvert whose big hole was visible. She would snort and shy, and without the slightest warning, jump several feet to one side or the other. This was more likely to unseat her rider than if she reared, and her "fright," we suspected, had this double intent. It behooved any rider to keep his or her eyes open for such trifles along the route because one was then at least partially prepared against being thrown from the saddle. The fact that Emmy's middle was so round that no saddle could be securely cinched for too long made this behavioral quirk particularly troublesome.

I do not remember Emmy's ever missing an opportunity to match wits with her human family, to say nothing of her own kind. Her always being so alert was probably one reason she assumed leadership of our small herd of horses after Chub's death. She regarded the initial process of being caught as a test of wills. When the horses were out in the pasture, which was almost always the case, Emmy had no intention of being interrupted in whatever activity she was engaged in at the moment, in order to be put to work. Hence, she easily avoided allowing anyone to come close enough to grab her. It had been a simple thing to catch Chub or Floss, who were always amenable to being approached, and once one took hold of the forelock or laid one's hand over their necks, they were as good as bridled. But Emmy would never be that completely broken to obedience. To grasp her forelock was an invitation for her to jerk back and break away. Perhaps early training had not been as expert as it should have been; for sure we were neither consistent nor regular enough in riding her. Being ridden was not a concession to superiority so far as she was concerned, but rather a condescension to a limited partnership in a mutual endeavor, with allowances to be made and accepted on *both* sides. My father, who had trained Emmy, never settled for this mutual arrangement stuff, and Emmy knew it was difficult if not impossible to thwart or disobey him, once he had caught her. But with me, she knew I could never push the contest to its limit.

Catching and bringing in the horses from any of their several pastures became a matter primarily of catching Emmy. Further, it became a matter of my catching her, for she was more ready to believe that I had tidbits for her than anyone else. If Emmy were successfully caught the rest of the herd followed, even as it had been during Chub's rule. But if she cantered off and refused to allow us to get near her, the rest became equally as unruly. Hence, because Emmy was the prime mover, all efforts were concentrated on getting the reins over her neck, the magic first step of the formula. To achieve this meant, upon finding the group, to call to Emmy, to show the paper sack in one hand, rattling it loudly, while holding the incriminating evidence of our intentions, the bridle, with the other hand behind our back. This was almost always a successful method. Emmy would reliably respond to the sack, but she might also grab her bite and then rear back, turn on her graceful legs and gallop off, her tail arched in triumph behind her. In which case we might as well give up for that time. But if she persevered in eating from our hand, we could gently and unobtrusively slip the reins under her neck and back over. It was simply that she regarded standing meekly in the pasture and allowing herself to be bridled without preliminaries as beneath her dignity and mettle.

Our problems were not over merely by getting a bridle on Emmy. After getting the horses to the barn, saddling and bridling them, we would bring out slickers which we carried on our all-day rides to take care of contingencies such as sudden mountain downpours. Long rides meant we also carried a lunch. The slickers had to be fastened over the back of the saddle and no one minded this operation but Emmy, who always stomped about as we tied the leather thongs around the raincoat. However, when the paper sack with lunch in it was brought to her side to be tied onto a front thong, she flared her nostrils, snorted, and shied away from me. Sometimes I had to mount her, holding the sack in my hand, and tie it on after I was seated. Then it ceased to be quite so objectionable. But to stand beside her and tie on that lunch sack—a sound and color and shape she was intimately familiar with because it looked exactly like the sacks containing tidbits which were standard offerings—this, for some strange reason, she would *not* permit.

One of her most unsettling idiosyncrasies was her inclination to shake, vigorously, at the most inopportune times. Often, but not always, just after the second cinching, when we were sure the saddle was finally tight enough to mount her, she would give a shake enormous enough to rattle her teeth. It rattled loose everything else, and necessitated a third cinching.

Or, at the completion of a long ride, when everyone was dog-tired, and one's defenses were not on alert, just before dismounting, Emmy would think to hasten matters by shaking. It felt like an earthquake let loose between one's legs, about nine on the Richter scale. The other horses seemed to be able to wait until they were rid of their saddles, when they would all enjoy a good shake and perhaps even a luxurious roll on the ground, but Emmy considered it more fun to shake up her rider!

Emmy had another sometime trick at the close of a ride, which occurred as she neared the home gate. She possessed normally, a beautifully smooth, fast walk, an easy trot and a rocking horse gallop. Her legs seemed to have springs in them, and she looked as dainty as a slender-legged, well-formed mare should, who is actually walking on her toes. But on these exceptional occasions she would suddenly convert her easy trot to a choppy, springless, stiff-legged pounding of her front feet that tossed her rider into the air, even though one gripped the pommel to hold on. It frequently landed a man, preferably a strange man, on the horn of the saddle, with excruciatingly painful results! My father could stop that little trick when he was on her, because she recognized him as her top authority. But his riding was too infrequent. Emmy was not unique in this unpleasant habit. Many horses are broken of that tendency either by trimming off the front hooves before shoeing so that they are tender, or by not shoeing the feet at all. This however was manifestly impractical in the rugged, stony, mountain country which was Emmy's home. Here the horses needed every bit of hoof they could grow to protect themselves, and to have tenderized her feet would have invited lameness.

In spite of this list of grievances, Emmy was fun to ride, a challenger who sharpened one's awareness while we rode. One had to be on the lookout, for instance, for the flowers that grew beside or anywhere near the trail. Because one of Emmy's chief gourmet delights was thistles, she could never resist the urge to leave the path and make straight for one of the prickly purple blossoms which her sharp eyes and nose had detected. To enjoy these questionable delicacies was one of the unexpected deviations Emmy insisted on making, and if I were riding her she knew she could get away with it, but she had to keep her own desires more under control when Dad rode her.

Yet, even with me, when the situation was critical she could be not only a no-nonsense person, but a dependable and supportive companion as well. She could and did rise to one occasion magnificently when called upon to test to the ultimate both her stamina and intelligence. My brother Bruce and I late one summer started on an all-day ride to reach the top of the Continental Divide, 11,000 feet in altitude, some fifteen or twenty miles away from our place which was only 8,500 feet. (The same destination Chub and his herd had reached when Emmy was very young.) We were going to follow in part the abandoned Denver and Salt Lake Railway course, which had climbed "over the hump" of the Divide from 1912 to 1928, when it was the highest standard gauge railroad in the world. The rails had in turn followed partially the route of an old toll stage road, established in the 1870's by John Quincy Adams Rollins, who in his turn had utilized an age-old Indian trail.

This latter route followed South Boulder canyon directly westward from Rollinsville, the town founded and christened by J. Q. A. Rollins. Mr. Rollins took his toll and stage some five miles due west along the canyon bottom, following the creek, then veered northwest to follow a branch creek. It was one of the dozens of trails in the area by which the indigenous Americans had crisscrossed the Divide of the Rocky Mountains for thousands of years, as they sought, eastward, winter pasture on the plains, and westward, the ample summer forage and hunting in the higher country west of the Divide.

It was always an exciting region to me, filled with human history which fortunately, up to that time, had not done irrevocable damage to the natural beauty. There was a profusion of wildflowers. Martens would scurry away, screaming loudly at our unexpected intrusion. The bell-like clarity of kinglets, and the Swainson and Hermit thrushes calling across the canyon from the depths or tops of hundred-foot spruce or lodgepole pines were among the most thrilling sounds I have ever heard. The pungent aromatic spice of the pines engulfed us at midday, and rising before us were the crags and majestic cliffs of the Divide itself. There were old logging roads, long unused, that disappeared intriguingly into the forest, the bare railroad right-of-way that cut a swathe along the mountainside of big timber, some rotting, wooden tie-sidings here and there beside it, and higher up were the crumbling snowsheds, all testifying to tremendous human effort and former intrusion into this wilderness. When we used to ride this country there was not another human being visible or within hearing all day long. We were alone in a world that belonged to itself, which had indeed returned to its own, and which I looked upon with awe and unadulterated joy.

On this particular ride I was anticipating our first view of Yankee Doodle Lake, a round, emerald gem set in a deep glacial hollow at the base of the final climb to the top. The railroad had taken its tracks right around the edge of the lake, and wound for several more miles, south and west, to come back north, far above the lake. There it entered a short tunnel called the Needle's Eye, to make its last run (all of this on four percent grade) to Corona or Rollins Pass. I have never been able to learn the Indian name. A pity, because Indian names for everything are always appropriately tuned to ecological significance or religious meanings.

The old stage-toll route had had a stop at the lake, where a log cabin still leaned tottering in the winds. It was buried by snow each winter, and seemed obviously not long for this world. Rollins' route went past the cabin, and then doubled back to the east where he had taken his coaches and freight wagons away from the lake, in the opposite direction from the one the railroad later took, making his grade around the shoulder of a large, almost treeless mountain. The tracks of the stage road were still visible, and I wanted very much to follow them and see how they came out, a thousand feet directly above the lake, crossing a sort of hogback on old riprapping built by Rollins, which was still marvelously preserved and solid looking.

My brother considered both the old stage route and the railroad grade much too long. He proposed our climbing straight up the precipitous incline that encased the lake, where we would meet the Rollins road and then continue to Corona. This would save between two to five miles, which in this high and rugged terrain could mean a good hour or three on horseback. It was then late in the morning, and even though we were at that point close to ten thousand feet in altitude, the sun was very hot, and the insects still numerous and quite virile. So Emmy and I followed Bruce riding his mare, Cassie. The slopes immediately above the lake were not too difficult, in fact were more like meadow land with flowers and bright green grass, such as grow only at the edges of snowbanks in summer or late spring. But about two or three hundred feet above, the going became much rougher, far more so than it had looked to be as we surveyed it from below at the edge of the lake. The horses were laboring, so we got off and prepared to lead them the rest of the way.

The view was increasingly magnificent and indeed startling, if one looked straight down into the very dark green depths of the lake. The horses began to slip on some of the looser rocks. A misstep could dislodge a stone and one's self as well, and once a rock started to roll there was absolutely nothing to stop it until it landed at the railroad track, if indeed it stopped there and did not careen on into the water. It was not only stones which could roll like that.

I suggested that we go back down and take the route I had wanted to. Bruce, who was nothing if not determined, argued that we were part way now and had better continue, or lose more valuable time. After my third request to go back he suggested that I do so if I wanted but he was going to the top. I knew I would have great difficulty managing Emmy, who was an accurate judge of my own lack of firmness; I knew that even if I got back down to the lake I could never make her go in an opposite direction from Cassie, nor to turn once more away from home, and follow a long trail around the mountain and up over the hogback. So, I followed after my brother and Cassie who by now was trembling, both from effort and fear. The rocks became boulders, and there was absolutely no place to walk between them. We stepped from boulder to boulder and hoped each was securely in its place. As these were merely huge blocks which had disintegrated and separated from the bedrock it meant that they could be ready to roll downward if their equilibrium were disturbed. Climbing straight up at that altitude is tremendously hard on breathing, especially if one has not been accustomed to it and is not in Olympic condition. The horses were not much more fit than I for they led a rather easy life—pasture and an occasional ride. Cassie's breath was coming in great heaves, her nostrils were distended, showing the red patch high up on the inside nasal wall. And her eyes rolled apprehensively.

Emmy? She appeared perversely to be enjoying herself! She seemed to regard all this as a challenge or a lark and showed a remarkable sympathy for our situation. I would leap onto a rock, holding her reins in my hand. And she would wait until I steadied myself and then she took a short hop and landed there with me. She would nuzzle me gently, as though to say, "Well, we made that one, didn't we?" Her breathing was not as labored as mine was, and she was most certainly not upset nor panicky. I looked up at Bruce and Cassie now some fifty feet above and a little to the side of us. Even that far away I could see Cassie's whole body trembling. She who showed no common-sense caution about entering mine tunnels with my brother, tunnels which Emmy had no trust in, was now the most frightened of all. Unless perhaps it was I.

Emmy seemed by contrast, though more delicate and more highly strung than the plebeian, rangy Cassie, to have been born to the test. Her Arabian lineage was showing: under fire she was magnificient, not only able to handle herself superbly but extending her strength to me, in the true tradition of the Arabian horse and his master.

When at last we clambered up and out onto the old stage riprap bench, my brother finally became concerned that he had perhaps pushed things too far. My face, he said later, was literally purple. To this day, forty years and more later, I can get up a sweat remembering the exertion of that climb, and even more, its dangers which we were lucky enough not to bring tumbling down upon ourselves. Of all four of us I think Emmy came out probably the best, all around.

We had not finished with our hazards yet, however. After negotiating the perils of that climb and arriving at the ridge above the lake, we remounted and followed the old railroad track. Close to the Divide itself, the road traversed the side of a mountain whose north slope slid downward for a couple of thousand feet almost perpendicularly. To cross one part it had been necessary for the railroad to bridge the gap, and this was done by two trestles, each about one hundred feet long. The road had been abandoned some two or three years before our initial ride over it on horseback, after the famous Moffat Tunnel had been bored through the core of the Rocky Mountain cordillera. The rails were still there, and the solid bed of rocks and ties between. But a foot to either side there was nothing, just air, and the horrendous view of an avalanche of sharp granitic rocks stretching away below. Further, as Emmy stepped upon the trestle, it gave back a hollow ring. She pricked up her ears, advanced a little more, and then as she looked over to her right, beholding the two-thousand-foot drop almost beneath her feet, she shied violently to the left. As I recovered my balance, she looked down that side, saw the same gaping vacuum, and shied again to the right. There was no room in which to dismount—on either side; nothing to do but to stroke her neck and soothe her, feeling none too reassured myself!

But she would not have been Emmy had she not very quickly regained her composure and her good, common sense. She continued to step gingerly concentrating on her footing. I refrained, understandably, from gazing downward too intently either to the left or to the right, and so we made it across without further panic. Subsequent crossings on later rides brought only the momentary reluctance, but no more shying. The initial experience had not been one of her shying games, put on as a show, to tease or to throw a rider, but a genuine fright to both of us. In justice to Cassie it must be noted that in spite of her tremulous fright during the climb out of Yankee Doodle she gathered herself together and negotiated the trestle with more dogged and resigned determination than highly strung Emmy.

When the chips were down on the perilous slopes of Yankee Doodle Lake, Emmy was calm, dependable, and thoroughly trustworthy. She and Cassie both exhibited as wide a range of temperament as their human companions did.

In due time, after Emmy was grown, when she was three or four years old, my father decided she should be bred. Her character, her physical build, even her gremlin tendencies were just right for the type of riding we did as a family. Blooded, high-strung racing horses or pacers did not belong in rough, rocky country, where patience, a goodly portion of common sense, sure-footed, burrolike instincts, were more valuable to both horse and owner. These characteristics needed in addition the fortitude to endure the cold, the hard winds and the snows of the higher Rockies. In varying degrees Emmy had all of them. We hoped her progeny might inherit them.

Dad had heard via the grapevine which trails around and in and out among ranchers, that a stallion of parts was available at a ranch in the foothills, some 25 miles away. So arrangements were made to board Emmy there for a few weeks. In the light of what transpired, it has to be admitted, in our defense, that we had no idea what Emmy's behavior would be, inasmuch as she had never before been away from home. I suppose, if we had thought about it, we could have surmised what would occur.

But no one had. Emmy, however, had her own ideas which I am not sure I can analyze. But it would seem likely that she suffered from an extreme case of homesickness, both for the Ranch as well as for the other horses whom she had always known. Further, and above all, she showed a deep-seated indignation against the methods which some men use in breeding. Mares, unlike cows, who seem much less discriminatory in their acceptance of a bull, would appear to have preferences, and in addition, to be violently opposed to the proximity of a stallion until they deem themselves ready.

Whatever may have been her reasons, Emmy reacted with loathing for the stallion who was placed in her stall, and she almost killed him. She furthermore made herself inaccessible to anyone on the ranch, for riding or for handling of any sort. After a couple of weeks her behavior clearly indicated her complete and total rejection of any and all the humans, the stallion, and every other horse on the place whom she drove away from food, water, or her mere vicinity. If she did not know it before, having been rather consistently indulged in her wilfulness, she soon learned the psychological value of bullying and bluff and the advantages of taking the offensive, especially with strangers.

Dad received a call from the rancher and owner of the stallion, who remarked with some asperity that he had better come and get "the Espy Horror!" Her other sobriquet had become "Diana," since she consistently preserved her chastity. We got in the car and went down, carrying saddle, blanket, and bridle, and a paper sack with blandishments for catching.

Since on this ill-fated mating venture she had become so obstreperous and impossible to handle, we wondered how she would react to us. Or if indeed she might be glad to see us. We spotted her grazing in a field by herself, as we came along the road. I called to her from the car. Emmy lifted her head, and looked intently in our direction, the tips of her ears pointed so earnestly they almost seemed to meet, as though she could not believe what she was hearing. We had our own private call for Emmy. In repeating her name, it had become fused and came out sounding something like "Mee-amee! Mee-amee!" It was more or less inimitable, and she of course knew it instantly. I kept calling and she continued looking for the barest moment. Then she broke into an excited trot, following along the fence. When we got to the ranch house, Emmy was standing at the gate of the pasture, her head over the bars. It proved unnecessary, that paper sack with its bread snacks!

The rancher came out, exchanged some remarks with Dad, while we went over to Emmy, making no effort to hide the bridle in our hands, and slipped the reins over her unprotesting neck. She even nickered low as she nuzzled my hands for the tidbit she knew would be there. Emmy very seldom deigned to nicker at us humans, and her speech to her own kind was more likely to be an imperious squealing demand than mere conversation or pleasantries. This was the only time I can rememer that she showed no inclination to play hard to get, even acquiescing to the bridle. Her family had come! She was not abandoned! That was all that was important.

Since catching Emmy had always been no small feat of judgment, guile, and temerity, it was no wonder that she employed all her tactics on the hapless rancher whose stallion she found so repulsive. The rancher's amazement was quite justified, therefore, when he saw me coming toward the house, with Emmy bridled and following cooperatively. I have often wondered if she were smiling inwardly. Her eyes did have a certain sparkle. The saddle and blanket were quickly brought from the car and she was made ready for the ride home. I don't recall her even puffing up prior to cinching. When she wanted to, she had a very fast, springy walk, and I think we covered the 25 miles or so in record time. This ended her adventure at that ranch, and there was never a mention by anyone on either side of trying another time, another season.

I forget now where or who the stallion was who finally won Emmy over. But she did become a mother for the one and only time in her life. Interestingly enough, she also became as solicitous and indulgent a mother as Patsy had been before her. Her general disposition suggested she might have displayed an indifference toward her offspring much as Cindy had shown. But Emmy proved to be, if anything, over-protective of Lady Evelyn (the rather silly, overblown name I had chosen for the little palomino-colored filly). Perhaps her new responsibility provided even more license to lord it over the rest of the horses, an opportunity on which she would have capitalized with relish. Emmy indulged Lady Evelyn, nursing her well into her second year, seeming not to feel its absurdity any more than she had when she demanded the same thing of Patsy. She and her daughter were inseparable friends.

For some unaccountable reason, my father decided to sell Lady Evelyn to a rancher some miles away. Emmy was distraught for a few days, and ran about whinnying pathetically, then seemed to accept her loss. But to confound matters, Lady Evelyn apparently was not so satisfied to be separated from her mother, her home, and the other horses who had been her companions, and she found her way back to the ranch.

Fate was not kind to Lady Evelyn. She got through one fence, and onto the railroad right-of-way, was trying to get through the second gate, to reach her mother and her other friends in the pasture on the other side, when a train hit and killed her. One can only wonder what Emmy's feelings were as her only child lay dying beyond the fence she could not cross through.

Whatever she was doing, with her own kind or with us, Emmy's gremlins and pixies colored her approach. They were the means by which she had learned to get what she wanted. And this technique never showed up better than on our annual ride to visit our friends, Aunt Hallie and Uncle Jim. Emmy referred to her own faultless mental map of the trail reminding her where to veer off the path and find the purple thistles that had grown in that spot last year, and where the sweetest of the timothy had grown tall and full for snatching as she went by. I am sure she remembered equally as well the long and rich foraging on high country grasses which she could enjoy for several hours following our arrival at the Jordan mine property.

The summers when we went on our all-day visit to the remote cabin home of Aunt Hallie and Uncle Jim were during the late twenties and early thirties. Aunt Hallie was one of the richest women I ever knew, though her worldly goods at her death would not have brought fifty dollars at a generous auction. She it was who first impressed me with the validity of another set of values by which to live. She and Uncle Jim lived at the head of Moon Gulch, a two- or three-mile long, steep little tributary that fed into South Boulder Creek, along which our Ranch and Ice Plant lay. While we shared Aunt Hallie's hospitable lunch, Emmy ranged far and wide over the hills within their fenced property, her bridle removed and hanging from the pommel of her saddle, and only her halter on her head to recall to her wayward inclinations the fact that she was not wholly free and in her own pasture.

Even though Emmy knew that her immediate reward on arrival was to be set at liberty to find the choice spots, she never could resist putting on a display as we approached the cabin. It was a steep pitch, the last one hundred feet or so, but Emmy galloped up in a final spurt and finished with a thundering climax. On her initial trip there, she had without hesitation followed Cindy, galloping through the first gate and on up to the excavated bench on which the cabin stood. We never allowed that again. This levelled area had been reinforced by a rock wall base. A narrow set of steps penetrated the wall and led to the yard immediately in front of the cabin. Here Aunt Hallie grew the most beautiful columbine I have ever seen, great long-stemmed, luscious, lavender-blue and white blossoms, rising out of their generous lacey foliage. Invariably, and perversely, Emmy tried in succeeding years to climb those steps again, and deposit us right at the door. Not that she was so punctilious about delivering her rider. She knew she was not allowed in that yard; therefore it became the one place she was going to go, especially since Cindy was always permitted to lope up the steps and take over the porch. But after the embarrassment of the first invasion, we were prepared for Emmy's little charade, and brought her to a protesting halt at the base of the rock wall. Acting as though she were the most fractious of racing bluebloods, she would flare her nostrils, stomp the ground, shake her head as we dismounted, and skitter back almost rearing as we slipped the bridle off her head and pulled the bit from her mouth.

When Emmy was about ten years old my mother's riding mare foaled a dark bay colt whom I named Kim. Though Kim's first years promised a stormy and incorrigible career, after he was gelded and then taken gently in hand by a new caretaker who was a genius with horses, he developed into a long-legged, rangy animal, dependable, canny, an earnest, cooperative plodder.

Emmy's opinion of stallions being what it was, Kim's gelding made it possible for the relationship that developed between them. She and Kim, moreover, became the only permanent horse contingent on the Ranch, and as such entered into a sort of symbiotic companionship. That is, Emmy did the demanding and Kim either acquiesed or ignored her, letting her tantrums roll off his lean back. He might have become a weak and yielding nonentity. But he did not, and was definitely his own man (what had been left to him). Emmy could push just so far before he would either walk off and leave her, or turn on her in a rare display of anger. He was the quiet, retiring gentleman, and it was perhaps because she was unable to get up a good fight with him which so exasperated Emmy sometimes.

It always seemed to me that Emmy was a completely liberated female (long before the current use of that term). She was self-assured, and uninhibited by rank, sex, or species. She was at the same time a very social creature and needed the companionship of the other horses on the Ranch, if only to dominate them.

My father had built at the Ice Plant/Ranch in 1937, a year-round home called the Lodge. During the winter of 1952-53, the children and I lived in it. Em and Kim were then the only horses there. When we grained them in the barn, during bad weather, the oats or meal were as usual put in the small feed boxes which were built in above the long hay manger. Emmy, though then an elderly mare of 26 years, would with very immature haste gobble down her grain and then move over to Kim's stall and push him away in order to eat his. Kim was one of those canny horses who never eat their grain all at once; he mixed it judiciously with mouthfuls of hay, all of which he chewed meditatively; therefore he always had grain still in his box. And since Kim was also a gentleman he gave way to Emmy's unseemly greed.

We soon learned to tie Emmy to her stall in order to solve the problem. We didn't have to tie Kim because he knew better than to trespass on Emmy's grain even if he had been inclined. Long after Emmy had finished bolting her meal and was impatient under her restraints, he would be contentedly nuzzling the final bits of grain around the feed box, trying to get his broad tongue into the corners, while wisps of hay stuck out of his mouth. It was really almost more than Emmy could stand!

During that winter in the mountains, my son Dale took Kim out of the meadow every day or so and rode him into town in the late afternoon to get the mail. Emmy put on a beautiful show of devotion and concern, whinnying at the top of her voice, galloping up and down the fence as far as she could see him. Then she returned to stand guard at the gate until they came back. As soon as Kim got inside the gate, however, she promptly squealed, nipped him on his neck or rump, then wheeled around to kick him. Poor fellow, he must have wondered in his deliberate way about that bewildering brown mare.

She had other quirks of behavior which were uniquely Emmy. Such as her own concept of a proper cow pony. Herding instincts she seemed to possess, from some member of her varied ancestry, but she added her own fillip, by freely nipping a reluctant cow on its rear. I suspected that this added incentive Emmy employed was not due to her desire for efficiency. It merely offered her an annoyance she could quite innocently inflict, in the line of duty, as it were.

In the years before my father had to dispose of the cattle because he could find no one reliable enough to be left in charge of them, the dispersing of the hay began a regular merry-go-round for Emmy. She began by chasing the cows off one by one from the piles spread for them. After scattering the cattle she took only a mouthful from each mound as she went along. Sometimes it was harder to rout her own kind, but Kim was a sure bet—he never argued and he ate so slowly that she could enjoy the game a couple of times over before he was through. There were probably choicer bits of hay in some forkfuls but what made them sweeter or more delicate to Emmy was the triumph of usurping it.

However, to her credit, when Ginger and Muffet had colts, Emmy never ran the colts off, and might even share her pile with the youngsters. Could she have been remembering her own lost daughter?

Rough or recalcitrant as she sometimes was with some adult humans whom she knew she could bluff, she was surprisingly and unexpectedly gentle and willing when my children came along. They learned the feel of Emmy's back before they learned the feel of the ground solidly under their own feet. As they grew into four- and five-year-olds, they rode her and Kim bareback unhesitatingly over the hills and I never feared for them. Emmy did still consider it only fair sport to see if she could brush off an unwary rider under a low pine bough. After one or two such episodes the children caught on to that one. I have never ceased to be grateful to Em and to Kim for giving my children such kindly introduction to the joys of horseback riding, even as I had been introduced to it by dear old Chub.

In the fall of 1953, Dad was without a caretaker of any kind for the Ice Plant and Ranch, and finally hired as a last resort a long, drawn-out hunk of inept, lazy slipshod irresponsibility. He was a brackish backwash of earlier immigrant German energy and efficiency. It was cruelly unfortunate that Emmy caught the flu or distemper or a case of pneumonia, or whatever was going round that winter. She was then near thirty, and had she not had a constitution basically strong and virile she would most certainly have died right off. The fact that she lived at all, sans drugs, sans shelter, sans graining, was miraculous. But I suffer every time I think of what must have been her confusion, sick for the first time ever in her life, her family nowhere to be found, and left so cruelly to stand againt the cold and the wind, shivering and aching. The slothful individual paid to care for the stock never even phoned Dad to ask for help with the ailing mare, and perhaps was so slovenly that he never even saw that she was ill! So Emmy had to try to make her way back from the shadows as best her constitution and grit could manage it. My father was furious when he next came up to the Plant and saw her condition, her once sleek sides showing every rib under her heavy winter coat, her nostrils oozing pus, her eyes draining, and the sunken areas above them betraying both her age and the suffering she had endured. Graining and medicines which Dad ordered were applied—perhaps—when the non-caretaker felt like it, and Emmy did pull through. But never again was she her plump self, nor her fur glossy. And she retained a residual low-grade nasal infection. She lost too, some of her ebullience, and her tyranny over Kim ceased to be so important to her any more, a sure sign of failing physical energy.

1954-55 found Emmy in really poor shape. She never had recovered her round, smooth lines. It seemed impossible that she could survive the rigors of another winter on the Ranch. And that year winter did arrive early, cold, and windy. One day in January, my father called me on the telephone to say that he had just returned from one of his weekly trips to the Ice Plant to check on everything, had found Emmy down on the ground, thin, laboring to breathe, her nose running, and unable to rise when he tried to get her up off the snow.

He had called Andy, who was now retired but still living in town, Andy who had been mid-wife at Emmy's birth, as well as her tormentor in her early months. Dad never in his life possessed a gun of any kind. Andy who did, came with his. All his crustiness over the years slipped for once and Andy said, "You know, this is pretty hard to do. . . ."

Emmy died within yards of the spot on which she had been born, having with small but notable exceptions, lived her entire life of 28 years on one ranch, belonging to one family.

This is in itself something of a record, for as I read one time, of all the animals man has domesticated, the horse has for several reasons known more heartbreaking transfers of ownership and masters, homes and loyalty than any other. So Emmy had an unusual opportunity, and it was indeed our rare privilege to watch her development, physical, mental, and social, from her first unsteady steps, to the prancing filly, to confident maturity, and then into old age whose debilitating restrictions after her illness must have irritated and frustrated her.

She, like all other life, has her immortality, because what Emmy was had its indelible effect on what I am, just as what I am has its effect for better or worse on those who know me. She was loveable and exasperating at the same time, a creature of great native intelligence and perception, whose environment gave her an assurance and confidence which permitted her personality to develop as it did. And her particular character, so different from Kim's, who likewise was born, raised and lived all his life on the place, and with the same family, was another lesson in the uniqueness of the genes that gather to produce each living creature.

I hope, in whatever reservoir of truth and experience which may exist in this mysterious universe, that Emmy and I find ourselves close by sometime.

Pitty Sing

Chapter Six

PITTY SING

In the late 1920's my father allowed himself the luxury of venturing into more extensive development of his mountain property, expanding from the ice business into farming and ranching. His original foray into the cattle business was the modest purchase of a Jersey cow, whose milk was intended to help out in reducing the cost of his commissary bill during the two seasons when he boarded the ice crews, at winter harvest and storage, and at the time of the summer shipment.

Bessie (the trite name which followed with her) was a beautiful blending of shades of brown and black. She had big and very soft brown eyes, made even larger by their edging of black. Her soft mouth was rimmed by white. Bessie was gentle, and very tame, had known much handling and was very tolerant about the milking procedure, which was more or less primitive in our arrangements. Bessie was bred when Dad bought her, and her all-black calf we named Toffins, with no particular justification that I can now recall. Toffins was not a noticeably pleasant individual, never permitted us to be familiar with her, and remained aloof all her life. Her sire had been a Holstein bull.

By this time, Dad had decided the Espy cattle would be a Shorthorn herd. There was still a fierce loyalty to certain breeds whose virtues stockmen enjoyed extolling, and an equally fierce pride in the purity of their stock. In our part of the country at that time, the superiority of the Hereford (White Face) was unquestioned. The Shorthorn was next in popularity, and for similar reasons. Both breeds were fast growers and easy keepers; put weight on quickly even out on the range; calved with relatively little trouble, and a very low percentage of loss of both cow and calf; were not flighty, nor prone to disease, both bulls and cows tractable and easy to handle

The White Face were desired for their uniformity, the purebred almost never deviating in regular white areas on bright bay. Shorthorns varied a good deal, from pure red to mottled, to an occasional all-white, so the effect was not uniform; but their short horns were a plus, because many times they did not have to be dehorned. They also gave a greater volume of rich milk than the Hereford. Our Shorthorn bull was a huge, mottled, red and white animal, with whom I never felt much in common; nor I feel sure, did he with me.

I am grateful that by the time Dad bought his Shorthorn bull, Bessie's calf, Toffins, was old enough to be bred, and that in the late spring of 1931, she produced her first offspring, a little all-red heifer whom I named Pitty Sing. This youngster was thus the combined result of many generations of cattle breeding in Europe and the United States, and she possessed many of the best traits of each. Her milk was Jersey rich and Holstein plentiful, her dispostion Jersey gentle and Shorthorn calm and dependable, her good conformation mostly adhering to the latter. She had the native intelligence of all three, and was of course, as is every life unit, her own self.

About the same week in that same spring, one of the Shorthorn cows gave birth to a heifer. The two little red calves, looking so alike, became close companions, preferring each other over the rest of the new calves. Somehow their friendship suggested the "three little maids" in the MIKADO of Gilbert and Sullivan, so I named the Shorthorn heifer Yum Yum. But in my case Pitty Sing became the heroine. Pitty Sing's shade of red differed subtly from Yum Yum's, having a tinge of her grandmother's brown in it, and her face was not as broad as Yum Yum's, but had the narrower nose of the Jersey. Otherwise, they were indistinguishable as small calves.

Maybe it was the engaging behavior of the two close half-sisters that drew me to them. They were so active in their jumpy, erratic calf fashion, always side by side, whether curled in a nap, or kicking up their heels and their tails in sheer exuberant good spirits, racing together in stiff-legged calf fashion or investigating some interesting object like a butterfly or a field mouse, or occasionally indulging in good natured butting contests. They were about as inseparable and compatible as our two ducks, Jemima and Mimer, some twenty years later. They were invariably humorous to me, more so than adult cattle, whose appearance I also find amusing. They had to grow into their enormous ears as Cindy grew to fit the size of her feet; Emmy, the length of her colt-legs. All calves have straight, furry pipes for legs, finished at the bottom by polished cloven hooves much too dainty to fit the rest of their character.

As I used to muse over the fascinating similarities and differences of the animals on the Ranch at Rollinsville, I absorbed my first lessons in comparative anatomy; and a beginning comprehension dawned of the inter-connections of all life. It seemed to me cows and horses were distinguishable, just in appearance alone, by the fact that all the attractive curves, the pleasing proportions of a horse's body became, on the unfortunate cow, angular bony protrusions and ungainly, ill-fitting flesh in the wrong places; that wherever horses were beautiful, cattle were not. No horse ever could look so comical, no matter what it did with its ears, as any cow or bull must look, at any time, no matter whether its ears are forward or back. And no horse, even emaciated creatures, ever had hip bones that jutted in the unflattering fashion of the bovine. Even the eyes of cattle are wide set, and seem given to staring sideways, unlike the set and shape and light of those in a good horse. One could phantasize (if one believed in instant creation), and imagine that the cow was created first, and the crudities of her structure made more sophisticated in the next try, by rounding off the edges, and smoothing the corners, moulding the splayed feet together in one graceful hoof, and so on. . . .

But I came to know that a cow's appearance may be quite deceptive. In spite of all the tinkering man has done with the anatomy of cattle, solely to produce beef and milk, there are genes lurking there still unharmed, which give cattle survival value by virtue of a sagacity and hardihood, and social organization, all of which are quite obvious when one cultivates their acquaintance.

As my father's herd of cattle grew, it was fascinating to watch the establishment of hierarchy, and in a pasture of cows and very young calves, to observe the arrangement, apparently mutually agreed upon (but how?) for one cow to take over, for a specified period, the baby sitting for eight or ten mothers, while they foraged elsewhere.

I was finishing my first year in college when Pitty Sing and Yum Yum were born. By the time we arrived for the summer, Pitty Sing and Yum Yum were already too big and strong to be forcibly restrained by me. The only method by which to earn their respect and trust, to make them enjoy my company, was to fondle and scratch them in desirable places, talk to them, and above all, to encourage their affection by offering pieces of bread and cookies.

It was a significant aspect of Pitty Sing's character that through the following long winter months she retained her inclination to be friendly, even affectionate. Our contacts were spotty to say the least, sometimes as far apart as a month or more. She obviously had learned to respond to her name, which I hoped she would do, as I had worked very hard during our time together in the summer, to impress it upon her. She spent that winter as well as future ones, pastured with the rest of the horses and cattle in the big meadow. On every trip to the hills, if we had not seen the horses and cows as we approached, I walked as soon as possible after our arrival, down through the stubble, drinking in the perfume of the willows by the stream, and calling, "Pitty Sing, Pitty Sing," even before I could see her. Maybe the horses, who always crowded around for their expected handout, gave her the idea, but Pitty Sing was quite capable of hatching a thought on her own. Whatever triggered her response, memory or example, she would lift her head and start immediately toward me, Yum Yum trailing behind.

Thus we kept our association alive and growing whenever I came to the Plant. During the first winter she and Yum Yum grew enormously long coats, and looked so shaggy I hardly knew them. Like the horses, some of whom even changed color, they were entirely unrecognizable, because even the shapes of their faces were altered by the long fur. Pitty Sing's normally long face became quite chubby.

With spring, they emerged into fat, wide-eyed yearlings, looking more like twins than half-sisters. Their tongues were a surprise to me. They had been smooth early in their lives, but now they were as rough as heavy duty sandpaper. One can get well acquainted with a cow's tongue, because when feeding them a tidbit they wrap their tongue around the morsel, or one can put the food clear inside their mouth without fear of having a hand bitten, as a horse may well do. The cow possesses no upper teeth in front, and therefore cannot bite. Pitty Sing's tongue rubbed against my skin as she curled it about my fingers.

I was quite saddened in the spring by the sudden death of Yum Yum, who was killed by a gang of two dogs on the prowl. Ever after, Pitty Sing seemed a lonely figure, identifying to an extent with the other cattle, but most often by herself. She neither invited nor repelled the other cows, but she did not seem to feel intimately bound to them, and never minded at all leaving them to be with us.

As Pitty Sing grew to maturity, her mixed heritage began to show up in a number of ways. The square build of her Shorthorn father was lengthened more to Jersey lines; her glossy, all-red coat retained its slightly browner tint, her ears tilted more at an angle than the pure Shorthorn, and her face lengthened and grew more slender than those of the rest of the herd. She was always gentle, as a small calf, a heifer, and throughout her mature years, not just with me, but also with Mr. V., the caretaker who worked for my father during most of Pitty Sing's life. But she early showed, however, that she regarded me as her special friend. Though she proved she knew her name by raising her head to listen if anyone called her, she would come only for me. As far as my voice could carry, if Pitty Sing heard, she started in my direction even if she could not see me. When she did spot me she would start to trot, and then maybe to run.

Moreover, she refused to reply to any corruption of her proper name. It was amusing to watch Mr. V. try to make Pitty Sing respond. She was the model of tractability when he had to handle her, but she refused to recognize his call, because he persisted in addressing her as "Pretty Thing." He thought her name was silly baby talk and he would not condescend to use it. So he doggedly continued with "Pretty Thing," and Pitty Sing just as stubbornly ignored him.

Through the winter months the horses and Pitty Sing were almost immediately aware whenever the family was "in residence." After the summer haying, the horses and the cattle were all pastured in the meadow. The county road ran along the north side of the canyon parallel to the meadow, and if the animals were visible as we drove up, my brother and I always called to them from the car. The horses would all raise their heads, and with one synchronized motion start for the gate, far to the west. Among the cattle there was only one head lifted, and only one detached itself and followed the horses. By the time we reached the Plant, the horses were pushing each other around to be in what they considered the best position at the gate, and Pitty Sing was standing at the rear, her big ears pricked in anticipation.

During the summer they each had their methods of knowing where we were at almost any time. Sometimes it would be only Pitty Sing who stood at the kitchen door, or under an open window of the dining room where she could hear our voices. If she got no response she would call to me, in a long, low "moo-oo-o." If the horses were not in the upper pasture but in the yard because we were going to ride, they would take up their stand by the kitchen door, and Emmy would demand the choice spot by the open window. They had been doing it for years, and were long-time experts at wheedling handouts.

When both the horses and Pitty Sing were waiting outside, it required a bit of daring to come out, armed with offerings, because manners were thrown to the winds. Not one of the animals was really hungry, but they acted as if they were starved, and there was a chorus of soft nickers from dilated nostrils. Their frequent displays of temper at each other were very likely to find me in the middle. It often required two people, one to take care of the horses, and one to head for Pitty Sing, who had been chased far to the rear, usually by Emmy. I always felt sorry for Pitty Sing on these occasions, relegated to the ignominious rank of an inferior, a hanger-on, so to speak. And so I tried to compensate with an extra morsel. She had been, meanwhile, mooing insistently ever since the door opened, stomping her feet in impatience; and her salivary glands were at work overtime in anticipation, requiring frequent swiping mop-ups. She reached out as I approached, her long pointed tongue wrapped itself around whatever I had been able to beg from the cook—biscuit or pancake or a leftover muffin—and Pitty Sing chewed it, her eyes half-closed in contemplation of her satisfactions.

It is quite true that Pitty Sing mooed a great deal more than any of our other cattle. I used to wonder about it, but now I am quite sure that animals who are much talked to, try to answer with whatever voice they have, no matter what the species. Pitty Sing certainly learned to. She could and did "moo" at me with great volume from afar, but she also developed variations on her one basic syllable, so soft sometimes, they seemed but audible breathing; at other times they were almost a "mew." She obviously listened to me, and during a lull in my speaking, would offer her answer, whatever it was. It was an endearing, deliberate, and very personal inter-communication. I could feel the vibrations in her long neck as she spoke, because I usually had my arms around it during our conversations. She seemed to enjoy this demonstration of affection because she would stand still for a long time. Emmy never permitted any such prolonged affair even under blandishments of handouts.

Of all the many human foods which Pitty Sing tried and enjoyed, by far her greatest favorite was chocolate. The variety of foods we offered over the years was not exactly common to bovine digestion but her health never gave any indication that we had harmed her. She munched on hard candy once in a great while, and not much oftener on peanuts out of the supply we kept for Rachel, the ground squirrel. Her most frequent goodies were bread, muffins, biscuits, cookies and hot cakes—all, contributions begged from left-overs at the dining table. But chocolate—it was like offering a gold nugget to a starving miner, or putting a slot machine with guaranteed jackpot before a confirmed gambler. I don't recall how the strange habit started, though it is likely I shared one of my Caracas bars with her once. Baker's semi-sweet chocolate used to be packaged, for fast consumption, in small tablets, individually wrapped in foil, a half dozen to a small cardboard box, with the Dutch Chocolate Lady pictured on the front. I frequently carried a box in my pocket, and on a horseback ride it was usually our lunch dessert.

Pitty Sing very soon grew quite excited over even the prospect of chocolate. She never actually stepped on me in her enthusiasm, though she sometimes came within half inches of my feet; nor did she knock me over, though she tried to poke her big mouth into my pockets and nudged and licked my hands in her unfeigned eagerness. She had excellent brakes, better than any horse, and from a fast trot or even a run, could come to a sudden stop, bracing all four feet.

Even if I crept up on her grazing in the meadow and she was unaware that I was anywhere around, she learned to recognize the mere rustle of tin foil as I purposely unwrapped the chocolate from a hundred feet away. She would raise her head, turn to catch the sound better, spot me, and start off at a fast trot, her heavy sides swinging to and fro, her ears as far forward as she could get them, her big eyes showing their whites, and her tongue beginning to hang out in anticipation. One had to have the chocolate already unwrapped or she would have taken it, foil and all.

Though Pitty Sing's craving for chocolate grew out of a chance discovery, her passionate effort to get and eat it was never a mere amusement, an end in itself to me. It was a means, an important means as it turned out, that brought us closer together, and we made it into a pleasant ritual that carried with it much fondling and talking. It created more companionship and understanding than we would have had without it, since it provided extended opportunity to study her and for her to know me. Jane Goodall achieved the same end of greater trust and certain familiarity which would never have developed if she had not introduced bananas into her study of chimpanzees. The way to the heart lies through the stomach in most species—not excluding humankind.

It was Pitty Sing who granted me one of the most deeply rewarding and exciting adventures of my life, and bestowed upon me the highest honor I have ever received. After sixty some years of an almost aching love for all things alive and responsive, I have come to believe that (aside from those with whom true communication has been possible among my own kind) the most driving need within me is to pass through the barriers of doubt, mistrust, and misunderstanding into that enveloping light which washes over me with its wordless beauty, when another animal says by its actions, "I trust you. I like you. We are part of each other."

Pitty Sing, in her third spring, as a two-year-old, was expecting her first calf. As a precaution she had been put alone in the upper pasture on the north side of the canyon, where there was good open forage, several fairly thick stands of ponderosa pine, and some willow growth along a rivulet as well as a sprinkling of aspen. When Mr. V. knew the calf was about due, he had walked over the area to try and find her but without success.

As it was her first-born, there was some apprehension about the delivery, and if she had already calved, concern that she should be brought in for milking as soon as possible. Milk cows have been bred to produce far more than their calves can consume, and if the bags are not stripped, especially following the first delivery, they very soon cease to produce the desired volume. So it was quite important that Pitty Sing be found, and Mr. V. was of the opinion that she had already had the calf and was hiding out with it, which would be a very natural thing for her to do.

I was delegated to find her, and to bring her and the calf in. Sometimes a cow who has stolen away to give birth and has hidden her calf has to be forced or beguiled to the barn. She will not only refuse to lead anyone to her baby but will try to mislead. Indeed, some of the gentlest cows have turned quite vicious when they have been discovered in their stolen hideaway. Those who are under the impression that cows are stupid should reflect on the cleverness with which some new mother will conceal her baby, and ponder as well on the immediate comprehension the baby must possess that he should lie motionless in his hidden bed until his mother returns to him.

The portion of our land where Pitty Sing had been put was a particular favorite of mine, a lovely, variegated paradise of birds and shadows and pungent pine scents, and beneath the aspens the spicy perfume of dark, moist and rich soil. No stark "delivery room" of steel and antiseptics could compare with such a setting as this!

I pondered on just what method would be best to use in finding her, making her understand what I wanted, getting her to do it, and trying to guess just what her reaction might be to whatever approach I chose. I thought about just wandering through the different wooded areas, but decided against that as I might startle her. So I walked just a little way into the pasture, chose a rock from which I could see pretty well all round me, and from where I also could be seen, and sat down to wait. I carried a brown paper bag with me, into which I had put some bread and muffins (no chocolate this time), enough to keep coaxing if she proved reluctant about accompanying me to the barn.

As I had not yet seen any sign of her, I began to call and to rattle the bag, the almost fail-proof means of getting a response. Before long, my eyes caught a slight motion among the aspens some hundred yards away. I thought I distinguished the familiar shape of a brownish red head with curving horns, and two great beautiful dark eyes peering at me from between the white trunks. But there was no answering "moo-oo." Should I advance? Or should I wait for her to take the initiative? Had I been Pitty Sing, it seemed to me, I would have preferred to make the decision, so I kept my seat and merely called her, shaking the bag every once in awhile. Would she remember we were friends, and trust me in this very significant and basic area of the strongest and most deeply ingrained, instinctive urges? To protect from all comers one's new born, vulnerable baby.

I have no idea how many minutes passed. I do know I was very excited. Then, there was more motion, the head advanced, became a whole body, and my Pitty Sing began to walk slowly out of the aspen grove, and right behind her wobbled a pure white calf, unsteady on its legs. I heard soft "moos" as Pitty Sing looked around every few steps and nudged him gently with her big mouth. She seemed not to hesitate, but came straight toward me, her whole bearing revealing full trust of me, paying me without words the highest compliment in her power.

Still remaining seated, I waited for her to come clear to the rock; then I opened the brown bag and offered Pitty Sing a muffin. She took it, but more as a matter of course, like a perfuntory greeting with something else on her mind than as a reward for having come when I called. We talked, she in short, soft grunts rather than actual "moos," and I patted the length of her soft nose, and scratched between her horns, running my hand down her dewlap. I hoped that this latter stroke might get me near the baby, who was showing interest in this new creature, the first of its kind he had seen. Pitty Sing seemed not to mind my reaching out my hand to her youngster, and merely observed with interest its tentative sniffing, followed by a typically sudden snorting attempt to butt it. She reached over and gave her small son a long swipe up his side with her rough tongue, almost upsetting his already precarious balance. It was a gesture of approval and reassurance.

I was so proud, and at the same time so full of gratitude to Pitty Sing for her frank and willing expression of confidence, that I hated to break the magic moment of communion. After treasuring it as long as I could, I finally rose and offered her another muffin, moving away as I did so, and coaxing her to come along. She seemed not unwilling and came down the hill, actually without demurring at all, walking slowly after me, both of us accommodating our pace to the uncertain wobbling of the little white son who kept close to her side.

If there is surprise at my method of bringing Pitty Sing down off the hill, the explanation is that unlike the rest of our cows, Pitty Sing refused adamantly and always, to be herded from behind. She would follow me readily, but regarded being pushed from behind as sheer indignity, a humiliation she would not accept. She was ever thus, from her early maturity on.

Even when remonstrating with me against being pushed from behind, Pitty Sing never offered to persuade with her horns. They were short, slightly curved horns, but one of them began to grow inward and pressed on her head, making it necessary to dehorn her. This loss made her resemble so much the other Shorthorn cows who had been de-horned I sometimes had to look twice to recognize her at a distance. It was also largely responsible for later tragedy.

Pitty Sing was very like our dogs in her diffidence toward any but members of our family. She would, if I were beside her, permit outsiders to pet her, but she was otherwise not at all inclined to accept attention from a stranger.

I would like to be able to end Pitty Sing's story here. But it is unfinished until I have poured out my heart in some sort of expiation which has been demanding satisfaction for forty years. It is beyond my power to atone, but in some mysterious way I feel sure that this very effort to confess will reach where I want it to, and some day, if there is more than ashes, we shall meet and I will hope to be forgiven.

There has never been a satisfactory explanation for what occurred; or any explanation at all, as a matter of fact. My father could not seem to analyze, or perhaps did not dare examine his motivation for what he did. Whatever the motivation, it happened.

My husband and I were expecting our first child in July 1940. Early that spring we had gone to Rollinsville for a weekend, staying at the Lodge, the winterized house my father had built across the gulch from Sunnybank. That particular weekend my father chose to ship some of the cattle to the Denver stockyards. I remained in the Lodge because I could not bear to be near the scene when those animals were loaded into the truck, personalities I had named and whom I knew, to be sent to that barbaric, cruel, and pitiless treatment which has characterized the commerical slaughter of meat animals in this country for a century or more. The stories I had heard and read of panic among animals being driven to the slaughter houses, when the stench of fear reached them from those already suffering unnecessarily crude, physical maltreatment before death, had infuriated and tortured me. It is unforgiveable that the human race should consume flesh which has not only given its life that we might eat, but has been forced to endure the most unfeeling and calloused abuse in the process. In subtle ways our society has paid the price for this evil doing and will continue to do so as long as we persist in such savage methods of killing.

There had been, indeed, early that spring a story on the radio and an article in a Denver paper about a panicky runaway steer from just such mistreatment. It had brought my feelings freshly to the surface, and imprinted even more vividly on my mind, the fright and bewilderment suffered by these hapless victims.

I had stepped outside the front door of the Lodge as the truck drove off way below me, with its burden of eight or ten head in the back, and heard the frantic bellowing of one cow above the disturbed lowing of the others. All the dehorned heads looked alike in the truck, so I could not identify one from the other, and, of course, had not actually wanted to know which ones out of the herd my father had chosen to send away. Nothing was said at the dinner table that day about the shipment, nor was it referred to later.

It was some weeks later before my husband and I were able to come again to the Plant for Sunday dinner. As always, I went quickly down to the meadow. My father and mother were already there, feeding the stock. I called as always, "Pitty Sing, Pitty Sing," and expected her to emerge from the willows as she had done so many times, her ears forward, her soft "moo" answering me. But there was no "moo," no red body trotting toward me. Perhaps she was with a group of cattle surrouding the hay truck. But she was not.

I called to my parents, "Is Pitty Sing here? She didn't come when I called her." Never before nor after did I see the look that came over my father's face. "She isn't here," he said in an odd voice.

"Where—" I began, and then all of a sudden I realized. "Oh, you didn't—" I screamed, putting my hands to my face, as though to shield the truth from hitting me.

"Yes—" again that odd voice.

The whole scene is seared into my memory forever. I turned tail and ran, I don't remember where.

My grief was and has remained so keen, my self-reproach is so sharp, that I can barely force my pen to set this down. It was my fault, in the final analysis; I did not recognize that the agonized cry for rescue came from Pitty Sing. It was she who had called to me from the back of that truck. My fault, therefore, that I did not scream at my father and his helpers and run down to make them unload my trusting follower who had always believed in me. I failed her, in her moment of greatest need, and thereby consigned her to a death of bewildering fear, of insulting ignominy, and wrenching, brutal pain—what a violation of the love she so trustingly and happily bore me!

It is the betrayal of that precious trust and belief which Pitty Sing suffered—she called and I did not answer—that I must someday make amends for.

Kerry

Chapter Seven

KERRY

Kerry offered my second intimate and sustained involvement with a dog. Because he came into my life when I was newly married and running my own household, he was the first for whose life and happiness I was entirely responsible.

As my husband was a struggling young doctor we had very little money with which to purchase a dog. So we turned to the most inexpensive yet dependable source in Denver, from which to obtain a good dog, the Denver Dumb Friends League. 1940 was still a time of financial hardship for many people. At the League's boarding kennels in May of 1940 was an Irish setter female whose litter of purebred pups her owner could not afford to feed. We picked out the reddest, longest-eared, and the most appealing of the eight or nine wriggling puppies and bought him—for seven dollars. American Kennel Club papers were likewise out of the question financially for the owner of the bitch to provide or for us to pay for. In fact, seven dollars represented for us at that time just about the money I spent to feed two of us (rather well) for a week.

We carried the six-week-old puppy home, cradled in my hands and resting on the generous abdominal shelf provided by our coming offspring. He snuggled down for warmth, seeming quite content. We spread an old coat of my husband's beside our bed, on which he curled up for the night. His transition from the companionship and comfort of his mother, his brothers and sisters, to new surroundings and people, was made about as smoothly as was Cindy's. So completely has the Irish setter become man's domesticated property that I suspect there is now something in his very genes which makes such leaps quite possible. Cindy's adjustment was part domestication and part canny comprehension. I consider Kerry's to have been wholly emotional, at that age anyway, which is not to denigrate his own brand of intelligence. He simply fitted into his new home and his human orientation in his own way. We searched for a name as Irish as we could find to fit his ancestry. He proved as well, to have the proverbial Irish sense of humor.

The rangy adolescence, which is humorous in all species, Kerry had in abundance. He was not only rangy but lean unto skinny, loose jointed, skimpy tailed, and otherwise growing too fast in some parts for the rest of his body to keep up harmoniously. Some of the beauty of a mature setter is the flowing "feathers," the long waving hair from legs and tail and underbelly. Of feathers he exhibited none during his first year. And I was almost afraid people would accuse us of not feeding him enough, his ribs were so prominent. His tail, though graceful in its final growth, was in his first year or so a wagging, somewhat pliable, unadorned stick, and he could without effort, in one wag, sweep everything off the coffee table. He learned, with his terrific sensitivity, to wag down low after one or two crashes. That he *was* healthy was plain to see, for he possessed that overflowing exuberance which is the happy sign of that mysterious force behind all of us

In addition to his extremely lean proportions, his limbs seemed for months to be hung from loose joints. When he lay down he threw himself in one confused heap on the floor, sorting out his legs later. With his intuition of what was permissible and what was not, he never, never climbed aboard the furniture unless he was ill (which was very seldom) and then I would find him lying on the studio couch, looking so forlorn and exaggeratedly mournful that I could not oust him.

One of his preferred resting places was in front of the living room couch. Wherever we lived later, that same couch was his favorite. The routine became a flopping-eared, loping sort of gallop through the length of the house, into the living room, to throw himself against the couch, then roll over, lying on his back, his head upside down, ears spread on the floor, with his long legs in the air but resting against the front of the couch. There he would doze off, waking himself after a dream of some great chase that had induced whimpers and twitching of feet. If not the dream first, then the inverted position of his head caused inevitably a most tremendous sneeze. He always looked like a person caught snoring, and after checking to see that no one was laughing at him, he sheepishly rolled over on his side.

Kerry's skeletal structure is much clearer in my memory than Cindy's, covered as she was by her heavier mat of fur; or than my mother's Tammy with his thick shaggy collie coat that quite disguised a rather small frame. Kerry's silky long red hair lay close to his body, accenting his leanness. His high boned skull with its prominent middle ridge, the very large liquid, dark brown eyes that were both deeply socketed and yet prominent, added to the general hungry effect. All his life, the close-lying silky fur made his every pound or lack of it, immediately noticeable. But as he matured, the gangly-lean became merely a long-lean body, streamlined and pleasingly proportioned, which he controlled with grace that lasted until his very late years.

Kerry was one of those individuals whose deep native intelligence, combined with an overwhelming sensitivity and capacity for love, made any Obedience Training quite superfluous, even as it would have been an insult to Cindy. Indeed, beyond a very modest collar, Kerry only once in his twelve years experienced the restraints of a chain or rope of any kind. Because there had been talk of enacting a leash law we tried one time, and only one time, to take him for a walk on a leash. But Kerry regarded it as a sign of reprimand which he could not tolerate. He was so humiliated that he simply lay down with his long ears close back to his head and the most imploring and beseeching look in his brown eyes. He absolutely refused to budge until that symbol of shame was removed.

When Kerry was about a year-and-a-half old we moved from the bungalow with its fenced backyard and the traditional sidewalks, where he and I strolled with Dale in his buggy. Our new Denver home had neither fence nor sidewalks. It did have an already extensive flower garden covering about half the back, which my husband cultivated with much loving and successful labor. Kerry had no instructions about the care and protection of flower gardens, there having been but a straggly minimum around the edges of the backyard he had known. He was delighted at the hours of companionship afforded by my husband's working in the garden. But his pleasure never exceeded the bounds either of the garden's edge or the narrow paths within it, where he might lie discreetly behind my husband's feet.

Cindy had been soundly scolded while still in Denver, because she found my grandmother's cherished baby-breath bush a cushiony bed on which to lie. And Kirk, our collie who joined the family after Kerry left us, was addicted to flattening the iris and a bed of lilies of the valley in our yard at 2049. But Kerry, all the days of his life, never trespassed on a flower bed or a garden, in the first or last of his four homes, nor on those of neighbors at each location. He just sensed with his uncanny depth of perception what was fitting.

In addition to our second home having no fence, we fronted on a venerable city parkway, tastefully landscaped with long stretches of green lawn, flowering bushes and large evergreens as well as deciduous trees. It was a delightful place to walk and one which Kerry enjoyed more than most of the human residents. Since Americans do not walk for sheer recreation and pleasure as do

Europeans, he very seldom encountered people on his jaunts. There were squirrels to amuse him and whom he amused, who were to be chased as far as he could, plus interesting odors known only to him. But I knew that within an hour, never longer, he would be back from his jaunt, lying on the front or back doorstep or asking to be let in.

During the four years we lived on the Parkway there was a family down the block which had a chow dog and whose car was a station wagon. In his wanderings Kerry had met up with the chow, and there was an immediate personality clash which no amount of familiarity dissipated, even as happens between humans. In fact, Kerry never developed any friendship with another dog, not even with the two who belonged to my mother and father, whom he saw rather frequently and under the most pleasant and inviting circumstances. But with the chow it was an active dislike, and there were a few altercations out of which, in spite of the chow's reputation, Kerry seemed to emerge on top.

Perhaps the chow figured he was getting even when he rode in the "family wagon," because he growled and barked ferociously from the safety of the open window as they drove by. Kerry was invariably infuriated and chased the car for blocks, growling and snarling with equal vehemence. What we did not realize at the time was that "chow" came to equal "station wagon" and the other way round. Kerry felt permanently obliged to chase and to challenge any station wagon or related model which went by our house. When we moved to 2049 in southeast Denver we lived a block or so from a dear friend and her husband who owned two cars. One was an ordinary sedan and the other, a station wagon. When Annabel and Ned went by our house in their sedan Kerry paid them no heed, even when they spoke to him. But whenever they drove by in their station wagon he came roaring out and chased and growled at the car as far as the corner. No amount of Annabel's speaking to him diminished his fury. All the years after his initial blood feud with that chow and so long as he could see and possessed enough energy he continued this particular grudge.

Though capable of unbounded love for his family Kerry was not a spineless milque-toast. He had his own self respect and boundaries which he considered decent and reasonable. He was three months old when Dale was born and not quite two years old when Laney came; thus he was an adult long before they were well into childhood. This may have given him a feeling of seniority; indeed, I was frequently chided by both Dale and Laney for being more protective of Kerry's rights than I was of theirs. They did learn very early to treat him with respect; and he found it therefore very hard to tolerate other children who wanted to maul him or to play roughly. He would retreat out of sight as quickly and quietly as he could make it, or he would come to my side, ears hanging low, his eyes mutely pleading, "Do I have to stand this? Can't you help me?"

His facial expressions reflected the gamut of all the emotions to which every animal is subject, from the abject misery of being told he had to stay home to the ecstasy of promised fun. He very early developed a smile at particularly happy moments. Drawing his lips back over his teeth as a dog does in a snarl, the more delighted he became the higher his lips went. Eventually something tickled his inner nostrils and made him sneeze violently. The more overjoyed he was the more he sneezed. We had to pretend we were laughing with joy and not at him because, like all our animals, Kerry possessed a terrific sense of dignity which could not endure ridicule.

In 1945 we moved to a spacious and elegant home some four miles or so from our Parkway house. It was an expensive facade for a crumbling marital relationship which dissolved within that year.... No outsider can ever know about the wounds which never heal nor those which have been sealed over with scars that do not fade away.... I mention the pain involved only because Kerry, as an extremely sensitive member of the family, endured all of it and because his life with me and the children was profoundly affected by the emotional turbulence, the readjustment, the change of life pattern, etc., which we all underwent.

I have often wondered if Kerry's only runaway (so to speak) was his reaction to the turmoil and disharmony which prevailed in the new and grander home. For not many weeks after we had moved in, and though he had an ample and beautiful park just across the street to roam and run in, Kerry chose to find his way back to our former home. There he took up his post on the front steps, refusing to leave. What a relief it was when the new owners phoned to report he was lying there. He did not repeat the incident. But was he returning to happier days and scenes?

Though he developed a generous area within which he timed his habitual roaming, he never ran away from his next and last home where I moved with him and the children in January of 1946. It was not so grand a place by any means, and things were not exactly joyful for some time to come. But there was harmony, which Kerry's responsive nature fully realized. There were several very difficult adjustments for him to make, chiefly the fact that I had to go to work outside our home for awhile.

I put the children during the day in a nursery school operated by the University where I worked. For Kerry this meant that he was left entirely alone for the first time in his life. He absolutely refused to allow me to go to work three blocks from 2049 without accompanying me, so I had to shut him in the house, a confinement he endured sadly and reluctantly. I would walk home at midday to let him out, but that was all he had in the way of companionship during the day, after six years of having shared our time totally. Weekends therefore became his joy, walking trips to the grocery, games in the backyard plus our long months of making meals into picnics in the backyard. We were the first family in the neighborhood to eat outside in the spring, and the last to go in, in the fall. It was our only form of recreation since we had no car, and almost no money, but it was a valuable and salutary experience in learning how much can be made of how little.

The years at 2049 brought Kerry and me together in ways that we might never have known had life not taken the course it did. His ready, silent gestures of understanding when I was troubled, a head pressed into my lap, a tail thumping on the floor when I smiled at him, or even a paw put on my arm in silent comprehension, his new insistence on sleeping by my bed immediately after I was alone, made me very reliant on his love and strength. I am not sure I met his needs with equally unfailing attention and perceptivity.

I spent many hours at the piano those years, playing out my heart and thoughts. Kerry would lie under the baby grand, or if I got too dramatic, he would shift to his favorite back rest against the front of the couch. I know he must have been acquainted with all the compositions I played, for my repertoire was rather limited, but he gave no indication that he was affected by one more than another—except for one piece, the "Meditation" from Massenet's *Thais.* To this particular piece he never failed to react. If he were outside the house he would come to the door, howl mournfully, and ask to be let in. If he were in the house he came immediately to the piano, and pushing himself between the bench and the keyboard, would raise his head under my hands, forcing my fingers off the keys. He simply could not let me finish. It became a sort of signature of Kerry's, his hallmark. What was the message that particular melody conveyed to him which he could not endure? After he was gone I could not for years bear either to play that piece or to listen to it.

In 1949 my father gave me a two-door Ford sedan which made it possible for us to journey on weekends to the Ice Plant. It was on these rides that Kerry found how he could unobtrusively take advantage of Dale and Laney to improve his lot. For safety's sake both children rode in the back seat and Kerry was stationed on the floor at their feet. At least Kerry started out on the floor but always arrived at our destination lying spread out on the seat, the children either on the floor or what remained of the back seat. How he wangled this none of us were ever certain. Like all children, Dale and Laney were inclined to sit on the edge of the seat. And Kerry apparently saw that he could quietly take over the vacated space. But once there, he sprawled so gradually that neither Dale nor Laney were aware of being pushed farther and farther toward the edge, and by the time they objected it was too late. Kerry was so miserable when he was scolded or humiliated and told to get down that it was easier just to let it go. Moreover, to stop at the side of the road for a showdown seemed like making a mountain out of a molehill. Hence he got away with it and once accomplished it was easy to repeat. He could look so very innocent, his head on his paws, rolling his big eyes and thumping his tail.

It was during these weekends in the mountains that Kerry made up for his confinement. He would run with all the pent-up energy and exuberance stored for days and when we took walks he would dash off to be gone a half hour or so, emerging out of the woods in exactly the same manner Cindy had, meeting us as though at a prearranged spot. A horseback ride would provide him with the same opportunities for extra forays, and he and Emmy early established an equitable relationship. He never tormented her by barking at her, and if he were careful to keep sufficiently far in front of her she never nipped him to hurry him along. His lean, dark red, silky form was so gracefully streamlined when he ran (except for the flopping ears), the long, sandy colored "feathers" waving and rippling from his legs and tail and underbelly.

Kerry was such a beautiful piece of life, so well-formed, so healthy and strong, so full of enthusiasms for whatever our days brought, and so eager to participate in them. Each year, in celebration of our first picnic of the spring in the backyard, he joyfully welcomed us to what was primarily his domain. He ran for his sock-rope and threw it in the air or worried it ferociously, so that it wound itself around his nose. He didn't mind how foolish he looked as he surveyed us, hoping for a tug of war game, with the ridiculous chain hanging down from his nose or around his neck. What a depth of wisdom we humans could put in storage if we were able to content ourselves with the pleasures of the moment. It is the secret of the joys of childhood, of the spontaneous play of animals, of their sensuous, immediate enjoyment of warmth. It does not diminish long range hopes and plans at all, it enriches the hours before we get to that distant goal.

Of such small things, though not all the same ones as mine, did Kerry construct his happy disposition. Totally reliant as he was on our companionship, our activities, our attitudes, he had to seek his satisfactions from us, as do all our domesticated fellow creatures. We have transplanted them from their natural milieu, taken them from their own kind, from attachments to their own species, and substituted ours, which is essentially alien to them. This has devolved serious responsibilities upon us, to make the lives of these creatures, in their forced dependence on us, meaningful and happy. They do not live by bread alone, any more than man.

In their long history of domestication, dogs have learned to create various territorial limits, to adapt their territorial imperative as they moved wherever their ever-restless masters took them. Kerry as an example had established around his various homes his "beats" which he patrolled regularly. He made himself at home at my parents' city house, whose rooms and floor plan he knew quite well. There he established as his favorite resting place my mother's most expensive Oriental throw rug, finding it, with aristocratic instinct, the softest and most pleasant place to lie. At Rollinsville, he discovered his own secret haunts to which he made visits whenever we were there. When one adds up these various environments it is quite remarkable, I think, that Kerry, as many other dogs have done, made himself at home in each of them. His lodestar, his constant, was our belongingness. Maybe it is man's scrambling of those ingrained and ineradicable territorial instincts which, by man's uprooting of those same needs, helps to assure the devotion for which dogs are particularly valued. They must seek assurance and security from man, who has overturned their natural instincts for his own benefit. I have often wondered what murmurings there must have been in Kerry's subconscious stirring him to an uneasiness.

It was Kerry's undeviating faithfulness to me which was responsible for his first heart attack, or, perhaps, heart strain. It happened under conditions which I would have avoided at all costs, if only I had had enough foresight. I blame myself for his suffering from that time on.

I was taken cross country skiing by a friend in the winter of 1947, when Kerry was seven years old. It was a modest trip and an even more modest beginning lesson for me. We invited Kerry to come along, which he was only too happy to do. Bill had chosen a gentle, open slope, in a valley of the front range. It was a lovely sight, the fresh snow having laid its burden on each clump of pine needles, a pale winter sun calling forth diamonds by the millions. Kerry was as delighted with the smells and the sharp fresh air as we were, and leaped around in the snow waiting for things to happen. When we began to enjoy the speed of skiing over the surface he started having difficulty in running after us, since each foot sank into the deep covering of snow. When we got to the foot of the run he reached us minutes after, panting as hard as I had ever seen him do during a horseback ride on the hottest summer day.

He regained his breath somewhat as we herringboned our way back up, because I had as much trouble managing my skis as he did managing his four sinking feet. The second run proved to be even more difficult for him and he met us even more slowly at the bottom, his breath laboring more and more. But such was his determination to keep as close as possible to me that he did not think of giving up and staying at the car. Instead we found him making only half runs, lying sprawled midway down to await our slow return to the top. His big eyes at last looked so pathetically haggard that I suggested we call it a day, and we packed up and returned home. Kerry was very, very quiet for the next two or three days, but I did not think much about it because our dogs had always recuperated in much the same manner from an exhausting horseback ride.

The next time, a few weeks later, when we got ready to go skiing again, Kerry was not jumping around wagging his tail in anticipation of being asked to "go to the mountains," a query that always put him in high gear. Instead he slunk away from us and lay down very deliberately at the back door, his head low on his paws. He came so reluctantly to me, tail dragging that I could not coax him further and left him home, for the first time in his life quite preferring to be there rather than with me.

I was troubled, recalling more particularly his actions, his obvious exhaustion during and following the last trip. But as the season for skiing ended about then, there was no opportunity for repetition and I to my shame forgot the incident.

Sometimes it takes absence and return to open one's eyes to conditions and/or situations which have been perfectly clear all along. On our homecoming from the very first separation we had ever known from Kerry, during Christmas of 1949, when we spent a month on the East Coast, I missed a heartbeat when I held his head in my hands and talked with him. There was a milkiness I could not dismiss in his once deep and clear brown eyes, and it seemed to me that he was straining to see me better. Or could it be his eagerness to reestablish security, to understand what I was saying? I stroked his noble head, with its high middle ridge of the skull bone, ran my hands over his lean flanks, felt his long silky fur of dark burnished red-brown.

Another of my regretted inadequacies in caring for Kerry was in the matter of his diet. I followed the advice of a family friend who was a veterinarian. Had I been a paying customer he might have given more thought to his directions, but I should have pursued a different course anyway. On discovering what I felt sure were cataracts forming in Kerry's eyes, I took him for examination and he confirmed my fears. But instead of suggesting vitamins and other ingredients which might have slowed their progress, he urged a strictly meat diet—horsemeat it was then. Years later when we took Kirk for needed dietary help, meat was considered exactly the wrong thing for older dogs. Thus I came to the sad realization that veterinarians, like all other doctors, "practice" their trade. And very little is really known yet, about good health in anyone. Kerry often did not want to eat the meat, which, it was suggested, I give to him in one large hunk so that his teeth would have exercise in chewing it up. What he did love, from the time he was a small pup, was a lamb bone. Frowned upon today, such table scraps have nourished uncounted generations of dogs, many of whom grew to respectable old age. But along with the growth of the cataracts, Kerry began to put on unaccustomed weight, his fur lost its sheen, and he developed almost a baldness near his rump. These symptoms I should have sought help for elsewhere, but did not.

Kerry, it was plain, was beginning to age when in September of 1951 my children brought home in a rickety shoe box a wounded mourning dove. Though he had never participated in the "sport" for which his breed had been cultivated, Kerry always did find birds of absorbing interest. This young dove, unable to fly because of her broken wing, was intriguingly available for investigation, her disability making it possible for Kerry to nose her and sniff her gently though she gave him to understand immediately that she would defend herself. Pecky was earthbound permanently because of her injury and hence lived mainly on the floor, quite accessible to Kerry.

Pecky found the forest of chair and table legs around the dining table a very suitable place to come and settle down for a nap after she had finished eating in the kitchen. Kerry, lying in front of the buffet not six feet away, would raise his head and observe her with ears pricked, then lay his head back on his paws, perhaps to watch her with continuing curiosity until he dozed off himself. Never once did he make any move to chase Pecky, to harm her, or even to startle her. He formulated his relationship with her solely on his own initiative, and without suggestion from us that he must accept and not harm her. His depth of understanding and his comprehension were so keen that I was able to leave the two animals alone together in the house when I left on an errand, confident that I would find Pecky as intact as when I had departed.

Kerry's blindness progressed of course. One of its pathetic manifestations was his inability now to catch the millers at which he had earlier been so adept. It was almost as painful to us in watching him as it must have been frustrating to him, to hear their annoying flutter and have to be guided only by their sound. Blinded by having to look straight at the light where the millers congregated, Kerry was pitifully incapable of his former skill.

How badly he was handicapped by his cataracts resulted in an almost disastrous climax one day in the mountains. The children and I went for a horseback ride late one summer afternoon, and, of course, Kerry went along. He did not now go dashing off into the woods for the long side trips he used to take, but kept fairly close to us, trotting either behind the three horses or leading the way.

As our dogs had always done, Kerry, after leading awhile, reversed his direction, and walked toward the rear of the single file column, as though counting to make sure everybody was present. We were going east on the trail, so this meant that Kerry was facing west and directly into the sun. Emmy was in the lead as always with Laney, Emmy the spirited, the fractious, the fast walker. I was next in line on Ginger, our flighty and clumsy blunderbuss who managed to stumble over things both real and imaginary. Dale was riding at the rear on Kim, the plodder, the patient one, the cautious.

No one could ever explain how it happened, but all of a sudden Kerry was entangled in Kim's front legs and feet. There was a short cry of pain from him, and then he was lying inert on the trail, with Kim looking down on him, his long stiff ears pointed forward, and his nostrils flared as he sniffed at the red body, so still. We all dismounted and knelt around Kerry, at a loss to know if we should move him or carry him. In a few minutes that seemed like an hour, Kerry did regain consciousness, got up slowly, and wobbled a bit. Then his strength seemed to return. He appeared not to have been injured anywhere, as he was able to walk normally if slowly. I walked with him and led Ginger home so that if anything happened I would be able immediately to avoid another horse stepping on him. If this accident told Kerry what it told us, he must have been confused and depressed. For it was clear that the sun had blinded him, due to the cataracts which were fast obscuring and dimming his sight. And we knew therefore in the future we had to be much more aware of this handicap and to use more forethought to keep him from getting into situations he could not handle.

Did Kerry know that he had helped to bring me through my most despondent and self-destructive years, and that he could relinquish his care, as I moved on to other phases? Whatever the reasons, conscious or otherwise, Kerry was fast slipping away the last months of his life, both physically and spiritually. He was more detached from his surroundings, perhaps because he could see less. Perhaps also because his heart at age twelve was showing the results of its overstraining some five years back, and he was very tired. At times he even seemed to be remote from me, far off for the first time in our lives together. Or could he have felt that Pecky's entrance into our family life had lessened the place he held in our hearts, and found his old age saddened by this intrusion and the reason for his hold on life less keen? I hope not, but I will never know in this life.

In August of 1952, both children went off to camp. There were only Kerry, Pecky and I in the household. One day during that time I accepted an invitation to attend my brother's birthday dinner at the other end of the city, leaving Kerry and Pecky free in the house together. While I was gone there was a very severe electric storm, something Kerry had come to fear with an almost irrational terror. When they occurred, he would run for my bed and dive under it, shaking and cowering for hours after it was over. He had also come to fear an open fire and gun shots, things that did not phase him in his youth.

When I returned that night, he was still under the bed, and in great distress, shivering, so frightened that he seemed almost not to know me. He did retain his house training, however, and managed to ask me to let him out. I did so, and regretted it almost immediately, for he disappeared and did not return, though I called and called. Not that night, nor all the next day, did he return. I went around the neighborhood frantically calling for him. The next night I awoke numerous times thinking I heard his low whine at the back door. But my hopeful "Kerry?" brought only silence—oh, such an aching, anxious silence. With each false alert I was increasingly distraught, fearful for his safety, dreading that he might have been run over and lay suffering somewhere alone.

About four the second morning I heard a sound I hoped again might be Kerry's whine. When I called to him my heart leaped at his responding cry. He *was* there! I do not remember how I got downstairs to the back door. Kerry was trembling and panting from fatigue and weakness—but he had come home! He wagged his tail feebly and sank down on the floor. I got him a pan of water which he drank greedily, though I was told later it was the worst thing I could have given him. He did however, close his eyes and seemed to sleep. I could not ask him to climb the stairs to the bedroom.

As soon as possible I called the veterinary who said he had probably had a heart attack and I should bring him to his office. Kerry could not rise and was uninterested in any food at all nor in any more water. My good neighbor helped me carry him to the car, where he lay on the front seat with his head in my lap, breathing painfully. I picked up my mother on the way and we stood by the doctor's table, as I held Kerry's head in my hands, and the vet eased his pain. He had made the last, supreme effort to be with me at the end, and died in my arms. No one could ever know how far he may have wandered in his pain, those two agonizing days, or how he managed to summon the strength to call to me at the back door. I am forever grateful to him, as his last act of love and devotion, that he *did* come back. He granted my heart the comfort of being with him when his own supporting and loving spirit left its tired physical body.

Kerry did not die, the elemental, unwavering certainty which was Kerry, shining through his eyes and illuminating every act in his life. I am as sure of this as I am of that same entity which is within me, projected on the days of my physical life. His supportive presence, his love which upheld me, his comprehension of what I said, and more, what I meant, brought me through some of the most difficult passages of my life. He kept my integrity intact, and permitted, nay, encouraged, the belief that my own worth could not be destroyed. "His vocabulary of comprehension was unbelievable," commented one of my close friends once. Kerry had become, I realized, forever inseparable from my own "being."

I learned that his unique imprint rested on all, when people in our neighborhood whom I had never even met, hearing of his death, called to say they missed the graceful red setter with the noble head and flowing feathers, whose self-assumed patrolling had kept the area cats in their place and the birds therefore more safe.

Kerry was the first of our animal family to be cremated. I have drawn much satisfaction, in an unashamedly primitive way, from keeping his ashes with me—some 27 years now. And when my own are scattered his will be mingled with mine, in a final appropriate physical blending. There will be lasting reunion on another level.

Pecky

Chapter Eight

PECKY

In searching for all the data we kept concerning Pecky, the mourning dove, who was part of our family for six years, I found more attempts to describe and analyze and evaluate our experiences with her than seemed necessary with any other member of the family. I had tried at least six different ways, and had abandoned each as inadequate. How could I convey the sense of wonder, the laughter, the anxieties, the uniqueness of our involvement in the life of a wild bird whose fate was accidentally wound into ours? Her impact upon us, the lessons she taught us, the love we experienced were so precious that no effort did or will produce it all quite as it should.

Our beginnings were inauspicious, a common occurrence. Two small children find an injured bird, bring it home, and everyone hopes it will survive.

In 1951, Dale and Laney were in the fifth and third grades. Riding their bicycles to school on the morning of September 16, they were startled by a grey bird darting in front of them, which, instead of flying off, ran for cover under a parked automobile. Realizing that something was wrong, they got off their bicycles and peered under the car. The bird did not move. Landy dove under, disregarding her clothes or hair. Unable to escape, the bird seemed immobilized and Laney emerged with it in her hand. She and Dale stashed her bicycle in the bushes and Laney walked the rest of the way, cradling in her hands the wounded little creature which struggled feebly to resist.

She pushed through the children who crowded around her and carried the bird straight to her teacher. Oblivious to her disheveled hair and soiled dress, and concerned only with the small life she held in her hands, she asked if she could have a box to put the bird in. The teacher, turning from the blackboard, saw only Laney's rumpled hair, her dirty dress, and her hands full of a mass of disarrayed grey feathers stained with spattered blood. She drew back with obvious distaste, saying, "Get that thing out of this room and go and wash your hands."

Shocked and dismayed, Laney resentfully went on to Dale's teacher who was more able to handle such emergencies, and a box was forthcoming.

As has happened, I dare say, to ninety-nine percent of homes with growing children, an injured bird was borne home at noon that day with great ceremony and care. The only difference with this small one percent was that Pecky lived. We will never know the miraculous, slender thread by which she was kept alive, both before she came to us and immediately afterward. What suffering and fear she had endured in the hours between her accident and the time the children found her, what additional and excruciating pain it must have been to be picked up and handled, even though gently, with those broken bones, jostled by all the motion of getting to school and then home. But there was no hospital to which a small bird could go, nor doctors to care for her, nor palliatives to be administered. Most veterinarians, unless specializing in their care, are loth to have anything to do with birds. Thirst and hunger must have increased her misery, her unremitting terror. But like other wild animals imprisoned or incapacitated by trap or injury, she could only watch, and wait, and hurt, and fear.

Her thirst we took care of at once. I pushed her head down into a small cup of water until it rose half way up her bill. When she realized that this was indeed water, she opened her bill and drank greedily, sucking the water in the manner of all the pigeon family, who do not gargle their water down as other birds. So dehydration was at least halted, but nourishment was something else.

We were fortunate in having our first inept efforts to care for Pecky aided by an ornithologist friend. He appeared at the house that first night, armed with splints and gauze to bind up the wing according to our description of her injury. It was such a comfort to see Lang take Peck firmly and with assured and knowing fingers, lift her wing and examine her. She pulled away, of course, and tried to peck at him, but a mourning dove in top shape does not have a peck which amounts to much. After examining the wound minutely he put her gently back in the box and said there was nothing he could do; the injury was in a place comparable to a break in our hand if it were bent at right angles to the wrist. There were too many fine bones involved, he said, and further, the wound was now too old. Lang prophesied that if she did live, she would probably not fly much, if ever. The break would fill if at all with an abnormal bony structure. He was quite right. No new feathers ever grew back to cover the wound; the bony growth protruded in a bare stub; the coverts, feathers which surrounded the area, came in every which way, creating cowlicks of a sort which she worked at all her life, and never could succeed in preening into the proper streamlined curve.

That there was damage to more than just the wing became

evident as she grew older. Her attempts at cooing showed increasingly as she aged that the left side of her breast was collapsed, that it simply did not puff out evenly with the right. Pecky was therefore more of a semi-invalid than we had at first thought. If she continued to suffer actual pain, she never gave voice to it. She accepted her handicap and lived with it in the framework of that marvelous adaptive philosophy I have observed in all animals. When they can no longer stand it, for whatever reason, they will to die.

Lang identified her as a mourning dove, probably one of the June hatch, as she already had her young adult plumage. This put her age at three months. Mourning doves in our area, he said, usually have two sets of young each season, one in June and one in late August. He said we would not know if she were male or female for at least a year, if the bird lived. He gave us some idea of her food requirements, and by experimentation we found what she liked best. We found a feed store which could supply the millet and brome and hemp. Pumpkin seeds, sunflower seeds, wheat and cracked corn were suggested by various people, but she turned up her nose at all of them, if a mourning dove could be said to turn up her nose! She did the equivalent anyway, as she impatiently brushed aside all that did not appeal to her. She did, moreover, have a remarkable ability to lift her upper beak in what very much resembled turning up her nose. The children dubbed it "noodle beaking."

By slow degrees and trial and error (through all of which it is a wonder she didn't die of malnutrition), we eventually gathered together the items of her diet on which she lived in comparative health throughout her life. She ate her favorite millet whole, never bothering to crack it as do the smaller juncos, siskins, etc. We bought washed gravel for her at a pet store which she picked over, eating only the grains about the size of a pin head. We learned through her own teaching that she should have finely crushed egg shell, which she ate sparingly except just before egg laying.

For the first few days, however, it was all a very temporary situation. We were never sure from night to the next morning when we came down first thing to see her if our small wild creature would still be alive. During those first days she lived in our basement, in a large box partially covered, on the floor of which, in the back, the children insisted on putting a soft piece of terry cloth. There were several reasons for my housing her downstairs. In the first place, she would be much less a constant source of curiosity and investigation on the part of Kerry, who after all, had an inborn interest in and concern with all things feathered, but who could easily be kept upstairs with us, and away from her, without scolding him and creating an unnecessary resentment.

Pecky's very first encounter with Kerry demonstrated that no matter how frightened, hungry and thirsty she was that day she arrived, and surely without much more stamina left on which to keep life flowing in her veins, she could muster ancestral instincts for survival. When Kerry came close to sniff her, she puffed herself out to twice her normal size, every feather seeming to stand on end. She actually hissed at him with her bill opened equidistant clear back to her double-hinged jaw, which changed her expression entirely. She then opened her beak wide and snapped quickly at Kerry's black nose.

By isolating her temporarily it also gave her time to recuperate and to mend if she were going to. Every time we came into the room she scuttled for the corner sheltered by the box lid, peering out at us with those enormous, soft eyes, whose mute pleading went straight to our hearts. Her peeking at us from her retreat caused the children to name her "Peek-a-boo." This gradually shortened and more appropriately became Peck, as she began to explore her new surroundings by pecking everything she saw. Thus, the strange alchemy which combines to produce most family nicknames.

Pecky at first did not eat when we were around. But we knew that by the scattered seeds, and the reduced water level in her drinking cup that she was at least tasting her food. The sand we supplied was considerably scratched over. We never expected her to survive past the temporary eating arrangements, the top of a peanut butter jar and the bright blue quarter-cup of a set of plastic measuring cups. This tableware became permanent, mainly because she became attached to it, but the bright blue of her measuring cup water dish came to mean water itself to her. We learned this to our interest and delight some weeks later, when she tried to drink from the dress I was wearing, which approximated the color of her cup. Attempting to interpret her actions, we produced her cup. She dove her bill in clear to the hilt, drank long and heartily, coming up for breath only when she had had enough.

Not to be overlooked, and I believe crucial to the whole process, was the love and tenderness which flowed out and into her every time any one of us came near her. She heard it in our voices, felt it in our handling, even though being handled so frightened her at first, going against all bird instincts. When it was necessary to clean her box one of us would hold her quietly, nestled in our hands, while the other did the cleaning and changing of papers. Pecky struggled at first, battling the enclosure of our fingers, biting as fiercely as her small strength permitted, but when nothing terrible seemed to happen if she sat quietly, she soon learned that it was simpler and less painful to take it easy.

From her darkened box it must have been difficult for Pecky to know just what was happening in the outside world, if indeed one still existed. Or did her instincts for timing, the weather, and all the other undiscovered and uncharted depths of a bird's knowledge and equipment tell her facts her human family could never realize she knew? Because Pecky was a wild bird as well as a migratory species, we felt we had to take her out after a couple of weeks and see if she *could* fly once more. Lang had predicted that should she survive (which he later told us he privately had thought most unlikely), he believed she would be earth-bound for the rest of her life. Now that she had endured what was probably the most critical period, with apparently no thought of giving up, there was an outside chance that she could manage flight. Moreover, it was the time of year when the doves were gathering to go south, and it was against the law to keep a wild, migratory bird captive. Dale and Laney were close to tears at the prospect that this small creature, already an important part of each day's excitement and interest, might leave us as suddenly as she had come into our lives.

When we first set her down in the yard, reintroducing her to openness, to the possibility of wild freedom, Pecky blinked and looked around her. New, yet ageless feelings must have surged through her small body. She vibrated noticeably, took a few steps through the grass, and crouched low to take off. The exhilaration of anticipated flight was obvious. She spread her wings. All the techniques of flying which she had mastered as a fledgling returned just as quickly. Her wings beat the air, though with more commotion than is normal, and she rose. But suddenly she was not going as she had obviously set her course, not even to the top of the fence. She put on more speed, and cleared the fence, but could not seem to guide herself. She kept veering to the left and finally fluttered down, tangled in the long branches of our neighbor's peony bush where she thrashed about helplessly and then suddenly became still. When Laney reached her, Pecky was breathing very hard indeed, and offered no resistance to being plucked out of her predicament.

It was now apparent to us and must have been to Pecky, that she could not and never would fly other than very short distances. Since it was also obvious that she had no intention of giving up on life we began to plan more permanent arrangements.

We thought it important for her to have sunshine and fresh air a few hours each day, so we set up a pen of chicken wire in the middle of the backyard where I could watch her easily from the kitchen. Though it was an unsubstantial enclosure she seemed to feel no urge to try and fly out of it.

This was a fascinating turn of events for Kerry who immediately took up his post beside the pen. He lay just outside the wire, his head on his paws, regarding her activity inside with mild but unflagging interest. This made Pecky exceedingly nervous at first though as time went on she seemed to accept him as an omnipresent and unavoidable drawback in her new environment. For his part we will never know if he considered himself her guardian when he would get up to scare off the sparrows who very soon spotted the food I brought out for her and came down in droves. As Kerry leaned over the top of the wire Pecky backed away in the direction of her dish and sort of fell into it where she proceeded to sit. I was not sure whether she was retreating from the sparrow invasion or avoiding Kerry.

As Pecky was now a member of the household we began deliberately to try and tame her, to accustom her to handling. At first she would run from us, but there being no place to escape to, she had to submit and found that it was not an insupportable evil. She rather quickly grew used to being held close to our faces and I am sure sensed the affection in our voices as we talked to her. What basic reassembling Pecky had to put together out of her normal inclinations, when she realized that her wings would never carry her as they were designed to do! How did she accommodate to the loss of her own kind? All of this we will never be able to fathom. But achieve them she did. Few birds in her situation do so, nor can do so, successfully. Other baby doves were from time to time brought to us who in no way accommodated.

For a number of weeks Pecky spent much time working at the gaping hole in her shoulder, which made my knees go weak to look at, pecking at it with that dogged patience and endurance of pain which animals invariably display when they try to mend their wounds. From the way her wing involuntarily pulled away from her beak it was clear how painful her attempts at self-surgery were. Before the dark hole filled with its bony growth she would occasionally extract a small bit of fractured bone. The growth not only filled the cavity but protruded in a stub that dug into her body when she did try to fly, making her hold the trembling wing out from her body after one of her inevitably unsuccessful attempts.

As a mourning dove Pecky had a naturally companionable nature. The two young hatched together in a mourning dove family start life together, and grow up exploring their world together. They are very soon left to their own devices by the parents who are more concerned with each other than they appear to be with their young, following the brief introductory period of feeding and instruction. The two youngsters eat and play and fly and drink together, they coo together, they sit close together and preen simultaneously, and their pattern is thus set for constant companionship with the opposite sex when they mature. Many ornithologists believe that mourning doves mate for life. Forlorn and lonely individuals of both sexes can be observed following the execrable open hunting season on doves in many states in late summer of each year, when the mate has been felled by a bullet, and the remainder of life is a truncated existence for the one who is left.

In spite of her isolation from her own kind, her aloneness and her unnatural living conditions, Pecky developed a remarkably vivid, if quiet, personality. She displayed likes and dislikes in fascinating revelations as she lost her fear of us. Our whole mutual experience was possible because of a fortunate combination of circumstances. Her wound was of a kind that could heal; she was a species whose food requirements could be met without impossible provision of exotic insects or grubs; she was young enough to be able to adapt and not mourn fatally for her kind or her normal wild environment, and at the same time was old enough that the loss of parentally provided food was not vital, which is so often the case.

The most meaningful element which made it possible for her to be content with us was the capacity she possessed to respond to our love. It was a most happy, if unlikely, pairing of a wild bird with a human family: her domestication, indeed her sophistication, with our resulting sensitization to the entire bird world. I have become convinced that perhaps the *most* crucial factor in the whole situation was *love,* and its power to wreak miracles. No other word seems adequate.

In a few weeks Pecky no longer ran from us. She even climbed deliberately aboard a finger pressed under her breast. Proof of her comfortable acceptance of her new life came on her own initiative. One evening some two months after she came to live with us we were entertaining guests in the living room. Pecky, by that time free to go wherever she wished in the house, appeared from a retreat under the dining room table, and in precise, small and ladylike steps, her feathers pulled down close to her large blood-red feet, she came directly toward me. Without preamble or hesitation she stepped up on the woven sandals I was wearing and proceeded to settle herself on one, fluffing out her feathers comfortably. It was a step forward in confidence of such magnitude, such an exhibition of spontaneous, freely offered trust and companionship that I have worn it in my heart ever afterward as a badge of honor. Later she preferred, when company was around, to sit in safety on the back of our couch, from where she was sometimes moved to hop onto our shoulders. On rare occasions she showed positive liking of a guest by hopping onto them likewise.

There were a few setbacks which tested her confidence in our benevolence, and one resulted from an early vestige of her former life. In the first weeks after she began to show that she wanted us to carry her about, we frequently saw or felt the crawling of an almost microscopic, transparent louse on Peck's feathers or on us. These parasites are not necessarily evidence of unclean habits; they are more or less endemic in the food chain of wild creatures. Pecky was meticulous about herself, but she did harbor other life which drew its sustenance from her. In return, when she found them she devoured the small mites even though she was not then and never did become much of a meat eater. She was moreover cooperative whenever we saw a louse and pointed to it, snapping it up and swallowing it, though she showed no interest in a fly or a spider likewise pointed out to her.

Whether we took the right steps to relieve her of these lice is a question, but there was no question about her indignant reaction to our method. I purchased a powder which was supposed to kill lice on small birds. Dale held the unsuspecting Pecky in his hands, and we turned her over on her back upside down. Now there is nothing birds resent more than to be turned on their backs, primarily because it renders them helpless. We sprinkled the powder into her feathers and under her wings, trying to reach as much of her skin as possible. It was a dusty ordeal and she sneezed along with us. Fortunately, one application of the powder proved to be sufficient. We never again saw a louse on her nor felt one on ourselves. Her hasty retreat afterward to the safety of the jungle of table and chair legs in the dining room and a period of pouting indicated that she suffered a relapse in her trust of us. She did occasionally act as if something bit or tickled her. As the children termed it, she got "the jabs," pecking furiously at herself; but as no louse ever surfaced we concluded these outbursts were no more than we exhibited at times with an "itch."

Topping even her disenchantment with us over the delousing incident was Pecky's later misadventure with a lemon meringue pie. After the sad revelation that Pecky could never fly normally, but along with the happy realization that she would survive under our care, we moved her upstairs, with her feed pen and her nighttime bed, a large cardboard box, into the kitchen. She took over, in stages, the rest of the house. She discovered early the pleasures of sunning in our living room, finding a patch of morning sun on the floor, stretching first her good wing then the maimed one, fanning her long, tapered, white-tipped tail.

Loving the sun as she did, and being earth or "floor" bound, it seemed only kind to lift her to the tiled window sills where she could enjoy more scenery and light. Dale and Laney thought the tiles cold, slippery and uninviting for Pecky, so they laid pieces of terry cloth on every sill in the house. Thus Pecky's love of sunning resulted in her spending many hours a day in one and another of the windows of our house, being lifted to them since she could not fly up. She did however sometimes flutter down if we failed to get her message that she wanted off. The kitchen window over the sink became a favorite as she chose more and more often to be where we were.

She was sitting on the kitchen sill on the evening of "the lemon pie." I had placed the newly baked pie to cool on my work table at the opposite side of the room, under which, on the floor, we had weeks before installed Pecky's feed pen. (This was her particular property, a wooden crate with a gate cut out of one side, where her water and seed, her gravel and mashed egg shell were kept in prescribed places. Such provision for habit means security in their environment and is essential to the well-being of all forms of life, I believe.)

Whether Pecky wanted to see what we were doing at the table, or whether she merely wanted to get to her feed pen, we could never know. We were startled to hear the noisy beat of her wings as she took off from the window, and before we could reach out to catch her in her erratic flight she was not only in our midst but in the middle of the lemon meringue pie! She sank, of course, right through the meringue into the lemon custard and thence into the crust beneath. Panicking, she flapped her wings to get out, pinioning herself as thoroughly as ever any fly did on sticky flypaper. The more she tried to flee the more she became gummed up until she was a solid mass of sticky goop with custard and meringue on her whole underside. The frantic flapping of her wings threw the contents of the pie all over her back and spattered everything in sight.

It was one of those first class messes, so awful one doesn't know whether first to halt the cause or stop the effect, or which end to start to clean up. The most obvious thing was to pluck Pecky out of the center of the pie. I fished her out with both hands, and carried the dripping, struggling, frightened little bird over to the sink where we turned on the warm water and the spray attachment. She was more and more indignant as the whole affair progressed—the accident, the mess, and the enforced bath. There was nothing to do but make her submit to her first shower which washed off the stuff on top, but to get at her bottom which was more thoroughly coated meant that we had to set her in a pan of warm water and sort of douse her up and down. Clearly one cannot scrub feathers as one can skin or even fur. (This is one of the tragedies of the thousands of birds mired in the infamous oil slicks to which they and we are now recurrently subject. You cannot scrub a feather!)

By stroking in the water the feathers under her breast and rump we tried to soak the custard and the meringue until it partially dissolved. Her feet were also gummed up with a little of the bottom pie crust added to them. I turned the spray under her, trying to reach the goop under her wings and in doing so released the pressure of my hand on her back. Pecky had had all she intended to endure and felt her chance to escape. She took off, spattering us all with much diluted lemon meringue pie, and flopping to the floor, literally ran as fast as her big red feet would carry her to her hideout under the dining room table. There she sulked the remainder of the evening while we worried that she might get a chill, for it was certainly not normal for a mourning dove to be soaking wet to the skin and at night to boot. We had not succeeded in getting all of the "sticky" off the feathers around her face. This apparently tickled and annoyed her for some days afterward for she spent much of her time scratching at them.

Kerry was an interested observer of the whole episode and felt it incumbent on him to follow and inspect Pecky after she had escaped from her bath. This added to her humiliation and reduced what dignity she had to nil. But by morning she seemed not to hold the disaster against us. Nor did she nurse a grudge against the pie tin in which I had held her to try and wash off her under-coating of goo.

For it became her bathtub, entirely by her own idea. Perhaps it does not seem an important detail, anything as routine as taking a bath, but under the artificial circumstances of Pecky's life a bath was another achievement in communications with us, in the development of her remarkable adaptive capabilities from the wild nonhuman bird world to the domesticated human world where she learned to live out her days in contentment.

Normal puddles after a rain, or a pool resulting from a garden hose, or other natural places which wild birds (those who take water baths) so often use were of course not available to Pecky. Since she so frequently sat on her "towelled" sill over the kitchen sink she often saw water being turned on just below her. One day she was pacing back and forth along the sill in an agitated manner. Then she ventured to step onto the slippery soap dish above the central spout and seemed to be trying to show me that water was her objective. I offered her a drink but when she tried to climb into the cup I divined her real intent and produced a pie tin, poured some lukewarm water in it, placed it on the drain board and lifted her to its edge. She stepped eagerly into the middle of the pan, plowed around in the water, swished her bill from side to side and lowered her head and breast into the water, wiggling her wings and getting her breast and neck very wet indeed. Then she stood irresolutely, half wet, half dry, seeming not to know how to go on from there. Eventually she climbed out and asked to be put back on the sill. It was obvious that to her it was an inconclusive and unsatisfactory performance, and I felt sorry for her as she preened what few feathers she had gotten wet.

Her efforts were so pathetic and inadequate no matter how enthusiastically she had begun that I tried to help by offering her a shower. I got her to stand on my finger while with the other hand I manipulated the spray faucet over her back and under her breast. If she were in the mood she stood very still, seeming to enjoy the sensation and even lifted a wing voluntarily so that I could direct the water under it. This was a very exciting development for me. When she had had enough she merely began to climb up my arm to get away from the water.

Seeing her lose her fear and integrate her living and personal habits in harmony with our human ones, we began to appreciate the evidences of Pecky's sharpness of observation, as well as her choices of things to investigate, some of which noticeably offended or pleased her.

It is not only domesticated animals who indulge in play. Pecky's detailed knowledge and use of her human environment was merely a transplant of inherited survival value, of the necessity to know one's habitat in the natural, wild state intimately and with a thoroughness "civilized" human beings have lost. Wild animals enjoy luxuriating for purely sensuous reasons in warm sun, or in a cool bath; they enjoy play—a game of tag or teasing one of their own or another species; they indulge in just plain curiosity, investigating some object never before encountered.

Pecky merely had more hours in which to amuse herself, since her basic needs were provided without much time or effort on her part. She never seemed to assert an ownership of any plaything as a dog will. She merely toyed with the object. But once she had exhausted its possibilities, she ignored it. Her playfulness also varied as does that of most household pets whose hours hang heavy on them. Some days she slept the time away and on others she was more wakeful, finding entertainment in whatever lay at hand.

One of the objects which she discovered early was a fuzzy, brown teddy bear of Laney's not much larger than she. For some reason it aroused very strong feelings in Pecky. Her round brown eyes becoming more oval in anger, she would run to peck it viciously whenever she came across it, aiming first for the black button eyes and then administering a more prolonged attack on the protruding little red felt tongue which eventually became quite limp and ragged under her assaults. Her jabs were meaningful and sharp, though actually not much more than a good pin prick in force. She was consistent about the teddy bear whether it was brought downstairs and put in her way, or whether she met him in the upstairs bedroom. He must have seemed a very real and powerful enemy in some way we could not divine.

A less threatening object was a marble-sized red pompom on the end of a white string which had come off Laney's bedroom slipper. The pompom was just a manageable size for Pecky to grab. She found it even more fun to bite the string close by the pompom and worry it as a dog would shaking it so furiously that the eye could not follow it; then she would flip it, sometimes landing it halfway across the room. Somehow we got in on a variation of this game. Presented with the pompom, she deliberately grabbed the string an inch or so away from the puff, shook it, and threw it. We retrieved it and then dragged the pompom slowly toward her. She watched it intently, then made a quick dart in its direction, grabbing it out of our hands. Her eyes during play of this sort reflected her pleasure and/or her mischief. But the interesting difference was that they retained their roundness, their soft light in contrast to the spark from her narrowed eyes when she vented her anger on teddy bear. Her span of attention lasted about five minutes before she was through with the game.

A pinwheel intrigued her for quite a while. She very quickly engineered its twirling by pecking at exactly the right point as we held it before her. She would watch with concentrated interest as it spun around and when it stopped she would peck it again to keep it going.

When I was working at my desk Pecky liked to be lifted to its flat top where she found numerous and varied things to investigate. Paper clips were a periodic source of fun for her. I kept them in a small box from which she would pluck them one by one, and either toss them wildly about or drop them carelessly around her. Clips later became one of the many odd materials she used in making the beginnings of nests.

The stiff brush on one end of my typewriter eraser was a source of curiosity to her and she occasionally picked it up and worried it. The typewriter itself was of interest also and she pecked at the keys, but only if I were using it. Since this was pre-electric typewriter days (for me at least), nothing much happened. She did not have sufficient strength or weight to force down a key. I think she might not have enjoyed the shock of its automatic return which an electric would have given her.

When I dialed a number on the telephone she followed by pecking at the big round holes. I could sometimes trace her progress across my desk by the clips, pencils, eraser, or even a small piece of paper with which she had amused herself en route. Sometimes, after exhausting her repertoire of playthings and if I were still working at the desk, she would settle herself on papers stacked in a wire basket and doze off.

The Christmas tree intrigued her and each year we lifted her into its branches before and after we trimmed it. It was, after all, the only tree she ever inhabited after her parental one was left behind. She obviously enjoyed sitting on it, settling herself down cozily. After it was trimmed she would look at the ornaments and give an exploratory peck or so at those near her.

Pecky shared mealtimes with us too. And not unexpectedly regarded some things on the table as of greater interest to her than others. For instance, she singled out our pepper grinder which was apparently just the right height for her handling. She would walk over the table to bite at the handle, knocking over the grinder. Once she appropriated a football-shaped, black vitamin pill about a half-inch long and found it amusing to roll around on the table. We finally set that pill on the top of the grinder. Pecky would march resolutely to it, lift it off, and proceed to chase it around the table. She never offered to touch any of the other food on the table except the butter, off of which she nipped bites at least once a day.

Sometimes it was neither convenient nor diplomatic (depending on whether we had guests whose understanding was not quite up to it) to have Pecky on the table. Hence we took to putting her on our shoulders from which vantage point she could watch what came to the table and decide what if any of our food she would like. The most delectable treat in her eyes became scrambled eggs, which she recognized, whether by sight or smell or both, as soon as the dish was brought to the table. She began to purr loudly and performed a drumming dance on our shoulders, in her eagerness pecking at our cheek to hurry it along. When she had had enough, she turned around and settled herself down with her back to the table and partially dozed off.

I was in the beginning concerned that if she survived and became a part of the household routine the children would, as children so often do, lose interest in her. This apprehension proved to be groundless and was less than fair to them, for their care of and fascination with Pecky as a special individual in the family circle never waned. She became instead the basis for a life-long absorbing interest in and interaction with the bird world. The minute observations which we made in our life together have helped us since to understand and appreciate all bird species so very much more. Her intelligence, the way she moved, her whole biology were uniquely available for us to study. Pecky prepared us for our life with the ducks, Jemima and Mimer, then with Wigga the pigeon, Gobi our Mynah bird, and even with Bee Bop the little hummingbird, each of which has exemplified the incredible differentiation among the species of birds. She taught us once again a lesson in the infinity of variation which makes all life so absorbingly beautiful.

Because of Pecky's availability we could study the astonishing variety of feathers which make up one bird's covering: the delicacy of pearl grey which colored the underside of her wings, contrasting with the darker, sparsely black-spotted, grey-brown of the upper, the deliciously soft and cuddly coating next to her slim body of what we called her "fuzzle feathers," that downy quality of the short thermal feathers so different from the stiff shafts of her long tail feathers with their webbing of long filamented barbs; each of the thousands of individuals shaded, spotted or iridescent and designed to fit into a perfectly functioning whole.

To examine a feather, for instance, is to realize one of the many marvelously intricate engineering designs by which Nature achieves her ends: the lightest possible construction combined with the greatest tensile power. We learned that it required about two weeks for feathers to be completely replaced. Black quills emerged out of the bare skin in the regularity of enlarged pores which we called her "porcupine" stage; next was the "paintbrush" stage, when a tuft pushed out of the tip of the growing quill; finally the protective sheath came off entirely and the feather spread into full blown maturity like a flower released from the enclosing calyx. Pecky assisted the final stage by pulling off small bits of sheath, which we called the "dandruff" stage for obvious reasons.

Pecky moulted at odd times and not according to the calendar which controlled her wild relatives. I have always felt I understood why hens are cross when they moult. Being covered with prickly little quills must not be very comfortable; further it must be a little chilly; and, finally, one's pride in appearance must suffer a very serious setback. Doves in the wild spend long periods combing themselves out. As a friend of mine said as she watched Pecky do her evening preen, "Well, she's put up all her pin curls now." All our birds have consumed hours in taking care of their feathers. I suspect if my body were covered as theirs with so many thousands of feathers of all sizes and kinds, I would spend a goodly portion of my waking hours smoothing, oiling, taking off sheaths, fluffing them out with a good shake, and then seeing to it that every one was back in its proper place. Especially if the hours of inactivity hung heavy on me, and if my family insisted on fondling and disarranging them, and most importantly because my genes insisted on as much feather perfection as I could achieve since life itself depended on their functioning properly.

That Pecky's feathers were always in good condition was a fairly accurate indicator of her health, and therefore that she was getting the proper food. Her droppings likewise were indistinguishable from those of the wild doves, easily cleaned up because they were rather dry and contained. (This was one of the reasons she could be allowed such freedom in our home and wherever we took her with us.)

There was no doubting that Pecky assumed even in her gentle, quiet, undemanding presence a dominant role in the functioning of our home. In fact, my concern over the possibility that the children would lose interst became rather that Pecky stood to suffer in their competition for her. We had finally to arrive at a compromise, where Laney's exclusive allotment was the first thing in the morning and Dale's the last at night. Laney's privilege was to go down to the kitchen where Pecky slept in her old cardboard box with the cover over it, that temporary arrangement which somehow had become permanent. She would bring her back to her bed riding either on her shoulder, in which case she had to come slowly because Pecky's perching and clinging ability was not all that secure, or she would carry her on a finger which was much easier for Pecky to grab.

Laney's bed was bathed in the early morning sunlight, and Pecky often enjoyed just basking in its warmth, snuggling herself into the long yarn tail strands of a big stuffed horse. Or, if she had awakened with more pep, she would teeter around the uncertain footing of the bed clothes, investigating by beak and tweak anything that took her fancy. At night, it was Dale's turn to take her to bed with him, and I retrieved her when I went down to say goodnight to him. She usually lowered herself in her slow, pneumatic fashion on his chest to sit quietly since she was tired by the end of a day. She would not, however, even though her hours were irregular and abnormal for a dove, go to bed until the house closed down for the night. She might doze on the arm of my chair or the back of the couch, or even on my knee—but be put to bed in her box? No. She would with a great racket push the top off and flutter to the floor, and walk clicking her claws over the bare floor to where the activity or the light was.

It was during the times she cuddled on the softness and warmth of Dale's knitted pajamas that we discovered one of the most interesting facts about Pecky. She purred—just like a cat. Sometimes she purred so hard that I could hear her across the room as I came in the bedroom door. If we put our ears to the mattress, the coils of the springs acted like an amplifier and gave the sound a metallic ring. She could, moreover, turn up or down the volume of her purring. It was vibration, a shaking throughout her body, and every bird we have had since—duck, pigeon, hummingbird, mynah—has done exactly the same thing, and for the same reasons, though not quite so loudly. If the rattlesnake, progenitor in the way of evolution to the Avian family, makes known its emotional reaction by this means, it seems to me not at all hard to make the connection. Pecky's purring was absolutely and purely emotion, and it denoted either pleasure or anger. It became one of her several successful means of communication with us. She purred in pleasure when she ate either at her feed dish or on our shoulder; we could both feel and hear it. She purred when we interfered with her wishes, leaning forward to peck at us crossly.

These discoveries about Pecky as a bird, finding out about her as her own individual person, plus the exciting revelations of her ability and willingness to adapt took place over the very first months of her life with us.

Near the close of the first semester, in early January of Pecky's first year in the family, I was asked to the usual parent-teacher conference which then was the custom in the Denver Public Schools. During our conversation, Laney's teacher remarked plaintively that she could not seem to get close to Laney at all. Knowing why, I reminded her of the incident of the injured bird early on in the school year and explained Laney's shock at the teacher's lack of sympathy for an injured and helpless creature, something she could not forget nor forgive. The teacher was overcome with remorse and tried in a praiseworthy effort to make amends by creating a special occasion for a visit by Pecky (now sufficiently domesticated that we felt it safe to undertake) to the combined classes of Dale and Laney. I kept the letters from the third and fifth grade students written after Pecky's appearance. Some of them revealed a new awareness, an awakened sensitivity to other creatures which just may have affected the way in which these children subsequently viewed their world. This awakening *is* the essence of Pecky's immortality just as it is the core of the immortality which we each inescapably ordain for better or worse in playing out our lives.

For this important day Dale and Laney divided topics to talk about—Pecky's playthings, her habits. She was carried around the room on their hands allowing those children who wished to, to stroke her or even to hold her on their finger. In the controlled situation it did not appear to upset or frighten Pecky or even to tire her unduly. We had no idea how she would respond and I was pleased and proud that she seemed to take this public performance in stride, and showed no fear, surprising for one so retiring and essentially so shy.

For the benefit of her audience she responded with her usual antipathy toward Fuzzy, the teddy bear, snapping at his eyes and tweaking his poor, bedraggled little red felt tongue on which she had so consistently vented her strange dislike. She snatched her little red pompom and tossed it. She preened and drank some water, even ate a few seeds and flipped more. She was a model performer, charmingly lacking in any affectations. Over the next several years Pecky made two more "public appearances," before a group of Cub Scouts, and her grandest as the guest of the Colorado Bird Club. She earned a fame far larger than the family circle in which she lived.

I suppose if we had not insisted on Pecky's being part of everything we did that she could merely have shared the roof over our heads and gone about her little habits neither noticed nor noticing. But I like to believe, and think I have reason to, that it was our constant companionship which opened up the channels of trust and its corollary, communication.

It seemed inexplicable to me that there was one item which she had refused to incorporate into her meals. All the animals I had ever known who ate grain or seed had relished corn. Pecky had consistently rejected the small cracked corn in her seed mix. At last she herself answered my question. Because one of her preferred perches was the window sill over the kitchen sink she became very much aware of all that went on in the cooking process. One day I was measuring some corn meal for a batch of cornbread when I became aware of her prancing back and forth. Then she hopped onto the soap dish. This routine was usually preparatory either to trying to fly off or it said she wanted me to give her a bath or a drink. At last I realized that she was very pointedly trying to reach the cup which held the corn meal. (Corn meal in those days was just that, coarser than flour and finer than grits.) Don't ask me how she knew she wanted it. Apparently her eyes, combined with that particularly acute sense of smell told her this was corn in the consistency she had been waiting for!

I offered her the cup and she gulped down whole billsful of the meal, purring loudly in her pleasure. Henceforth we provided a small jar top on the kitchen window sill which she emptied with regularity. She even learned to tell us when she wanted a new supply, stretching her long thin dove neck out over the sill when we opened the cupboard next to the sink where she knew the corn meal was kept. Occasionally she got so excited in her anticipation she would lose her footing and fall into the sink. This embarrassing accident we never commented upon, but picked her out of her dilemma without laughing at her.

The most remarkable part of the corn episode, as it was of her first bath, was the victory Pecky achieved in getting what she wanted by deliberately trying to communicate, by her initiating actions to get what she desired. Many humans are reluctant to admit that other species have the ability to think or reason or make decisions—even on a rudimentary basis. This seems to be among the last strongholds of the "me-only" syndrome which has discouraged if not prevented man from an open-minded study and appreciation of his fellow creatures. (The idea that only humans can see color is now finally out of the window. Tool making or using is no longer our unique prize.)

It is an abiding hope of mine that the time may come when mankind shall judge itself with sufficient, real humility that we are able to reach out the hand of fellowship to the other forms of life on this planet and see that, though they may not use our methods of conceptual thought, they *do* have concepts and are quite capable of making logical choices. And why should not Pecky have shared in this universal intelligence? From the situation in which she found herself, of having to adapt to an alien, unnatural environment, why should she not have used her brain to assert and fill her needs and reestablish her "inner certainty"?

She had come a long way in her own self confidence, as well as in her appraisal of us, when she could take the initiative in expressing affection, sometimes pecking gently at a mole or freckle, almost like a kiss. She had early shown acceptance, but the sharing of my afternoon nap was a great leap forward into fondness. And it was her own idea. As we have learned with our succeeding birds, affections must be given and accepted on their terms, i.e., when and how long they feel like it. Otherwise it is avoided, resisted or actively rebuffed. Pecky's "mushy" time developed at the afternoon nap. (Mimer's was immediately on our meeting in the morning, and Gobi's is the last few minutes as I start to cover her up for the night.)

Luncheon over, the children back at school and the house quiet, I took a half hour to lie on the couch at the far corner of the living room. Usually the two pets had already begun their afternoon nap. Pecky would be sitting in the patch of sun under the dining room table, and Kerry lying asleep against the buffet in the middle of the room. As soon as I was settled, Pecky began her pattern. She would get up, leisurely stretch out her good wing then attempt to do so with her deformed one, ending by lifting both wings high into the air, revealing the exquisite blue-grey feathers underneath. Then she would glide under the chairs, and walk deliberately and sedately across the room, her red feet barely showing under her body. Arriving at the couch she half jumped, half flew up, then walked over me deciding either to settle down on my leg or at my neck. If on my leg, a word from me began her audible purring and maybe even the wiggle of one wing. If at my neck she was so pleased with herself that she began to purr immediately. This tickled so that I could not bear it. If I spoke, then she vibrated so intensely it was like having an electric massage. She was offended if I shoved her away so I would gently put her down to my leg. There she would remain as long as I slept, dozing also. But the slightest movement on my part started the purring again.

One afternoon I had visible proof that Pecky could consider alternatives with whatever conceptual equipment functioned in her small, smooth, grey head with its observant, bright black eyes. I had just lain down on the couch and was aware of small movement under the dining room table. Pecky was on her way. She came straight under the table and the chair beside it, then stopped short as she reached the open space between us. She was looking at Kerry who was lying in his usual spot in front of the buffet alongside her intended route. His head was on his paws and his eyes were open, but he had not moved a muscle. After a short hesitation during which she was closely observing Kerry and assessing the situation she went back under the table, turned and came out the end farthest from him. Then, hugging the far wall of the room, she walked across its length, under the piano, and over to the couch, the most circuitous route she could have chosen as opposed to the direct one usually taken. Once at the couch all proceeded normally: she hopped up, walked onto my neck and settled herself down with great, purring satisfaction.

This performance displayed not only her clear decision-making but also her assessment of Kerry and his reliability so far as she was concerned. Her *modus vivendi* was a guarded working truce. I had left them together in the same room unsupervised numerous times; Pecky did not run from the big dog whenever or wherever she met him, inside the house or out, but she never completely let down her defenses either.

Her relationship to us grew from adapting in order to live, to understanding what was going on around her, soon after to comprehending what we said to her—or at least the gist of it—then to cooperating with us. She stepped onto our fingers quite willingly when we pressed a finger under her breast, and she stepped off when we suggested—unless she decided it was more to her liking to remain. She would dutifully cock her head to look at an object, then peck at a string, a needle, a pencil, a feather, or anything shown to her. But unless she saw some interest in it for her, she pursued it only to show us that she knew what we asked. Laney taught her to shake hands. For a bird this is something more of a balancing act than for a dog who still has three legs left to stand on. Laney would take hold of Pecky's foot and slowly move it up and down. Peck hung on and thus achieved "shaking hands," though she never got over her reluctance about trying to balance on one thin, red leg. She also learned to climb stairs. Laney began the lessons by coaxing Pecky to hop up a step or two at a time. As Pecky had become acculturated (or resigned?) to accomplishing most necessities on foot, she very soon mastered the ten steps to the bathroom landing and then in a couple of days the climb of four steps more to the bedroom level. After discovering this new ability she would often hop step by step either up or down from one floor to the other in answer to our call.

As is so often true of animals (including humans) Pecky showed some of her most endearing traits when she and I were alone together. There seems to be a lowering of the barriers of self-consciousness, shyness, timidity, or even exhibitionism which frequently make behavior quite different in the presence of more than two people than it is when the one-to-one relationship is unencumbered.

My devastation in 1952, following the loss of my loving morale builder, Kerry, plus some other emotional trauma, resulted in our spending the winter of 1952-53 in Rollinsville at the Lodge. The children went to school in the one-room schoolhouse in town and while they were gone all day Pecky and I had hours of unbroken isolation during which we intensified and extended our interpersonal communication.

She usually spent her mornings following the pattern of sunlight as it slid along the floor at the southern side of the big living room in our family lodge. She looked very small indeed, a little grey blob in the middle of a bare-floored, empty corner of the room. Whenever I walked by I stopped to speak to her. She responded more and more frequently by a slight flutter of the wing nearest me. This answer (whatever it was) would grow into a full size wiggle. Sometimes she would look around at her wing as though in surprise when it made a scraping noise on the floor. Other times she would stand up, get herself into position, stretch her neck and "coo." This was an emphatic, unmistakably positive response to me. She was never able to achieve a normal, full-throated healthy "coo." Because of the injury to her lung it was always a cracked, hoarse approximation. But even her imperfect attempt had a quality of ventriloquism which the normal wild dove possesses.

She continued her conversational purring at nap time that winter in the mountains and on another couch. There she appropriated the multi-colored, crocheted afghan which I usually spread over me. She seemed to love it—was it for its texture, its colors, its warmth? I never knew. She would jab at different colored blocks and settle down with the most obvious pleasure into its folds.

I often wondered if Pecky consciously chose the softness of some things because that quality was enjoyable to her. I am inclined to think she did, for this deliberation was no different from the sensuous satisfactions I have noticed other animals derive following the inclinations of their particular and highly developed senses. She found pleasure so often in cuddling on or into things soft, like the yarn tail of Laney's big stuffed horse, the settling into my embroidery floss, her preference for the soft pile of Oriental rugs at my parents' home in Denver. Indeed, from the time of her first visit there she amused herself by going busily on her own from room to room, hopping eventually down the two steps to the recessed living room and taking up her chosen position on the best Sarouk, that same one upon which Kerry always had chosen to lie when he could slip there unobserved.

One of her choicest spots at the Lodge became the long mantel extending across the big stone fireplace in the living room. Almost as grey as the rock, she would walk carefully from one end of the mantel to the other and settle herself down along the edge from which perch she had the advantage of being able to see wherever we went. If a fire were burning in the fireplace below her she sat with her breast leaning over the edge as though to warm it. We decided this was indeed the case because after a few minutes she would turn around and present her tail end to be warmed.

Pecky discovered a new game that winter which she continued to play whenever the proper circumstances came together. I was embroidering a pair of pillowcases and kept the various flosses I was using in a box which had once held a ream of paper. Pecky found it one day as she walked about the table on which the box sat. She jumped in without hesitating and with quick, decisive movements picked up the floss by strands and pulled it all around her, scratching with her feet like a hen making a hollow in which to lay her egg. Then with several small contented flips of both wings she settled herself into the slight depression from which she regarded me with pleasure plainly shining in her bright black eyes. She did create a deal of mischief every time she played in the floss for she accomplished a veritable rat's nest which had to be untangled. She enjoyed taking a jab or two at my thimble too, especially if it were on my finger. All our birds have recognized the alien look of a thimble on my hand and have found it necessary to peck it with varying degrees of disapproval. Ever after when I finished whatever embroidery I happened to be doing, Pecky was also finished with the floss. She would not be persuaded to show any more interest in it under any conditions until I took up my next project.

All of Pecky's activity was not limited to indoors at the Lodge. As we had when we were in the backyard in Denver we took her with us riding on a finger when we sat on the hillside or picnicked up the gulch. On her first free forays at 2049 in the spring of 1952, we had sat quietly in order to let her follow her own desires, at first tentative and hesitant. She found it pleasant to explore the yard picking up small blades of grass or eating a speck of dirt.

We felt rewarded when Pecky invariably returned to us after her preliminary wanderings, stepping carefully through city grass or hillside flowers to let herself down with deliberate assurance in the slow, pneumatic dove-fashion right beside us. Then she might decide to fan out a wing and her tail to catch a sun bath or she might compose herself, "dump" or "squash", for a nap. The children described her poses as "dump", a sort of rounded huddle, feathers fluffed out and head drawn in; "squash" was flattened out, the small head extended and wings pulled under her body very like a human lying on his or her stomach, with arms straight under the body.

She dearly loved a good sunning and sometimes when we put her down on the ground she would run as if called, to squat flat and spread as many of her feathers as possible for the sun to warm. Then the sun would make her skin itch and she would get up after a few minutes to scratch and preen assiduously. We discovered from watching Pecky that birds set their "scratcher" in motion, presenting head and neck to the rapidly moving claw rather than the claw moving to scratch different spots.

On the mountain side one day, quite before we were aware of it, Pecky spied a hawk lazily soaring high over the canyon. Her instincts rallied on the instant. She pulled her feathers tightly to her body, reducing her size of a moment before by almost half. She cooed a (for her) loud, scratchy, scolding squawk, and then, totally forgetting that she was earthbound, tried to take off the ground. Her inability only frightened and confused her further. We grabbed her and held her close, trying to calm her fears. But she was some time coming back to our "civilization," and I don't know if our intervention reassured her or not. The next time we sat somewhere near the same spot she acted immediately as though she had either just seen a hawk again or knew he would reappear.

Her comfortable habit of remaining near us on the hillside made us too confident, and our lack of watchfulness once almost frightened us out of our wits. We were enjoying in late spring one of our favorite breakfasts cooked over an open fire at our picnic spot up Willow Gulch. Pecky had shared in the scrambled eggs, walked around picking up small grains of sand, and found a spot to sun herself while we finished eating. Then we were preoccupied with the details of putting out the fire, gathering the dirty pans and dishes and packing everything to carry back to the Lodge, a distance of an eighth of a mile or so. We looked around to gather Pecky also. But she was nowhere to be seen!

Willow Gulch was aptly named for its thick and impenetrable growth of willows along the small stream which at that season was flowing in hurried ripples. Above the willows the land on both sides rose rather steeply, on the east a rocky, sandy area which supported a majestic open stand of mature ponderosa pines. The west side, on the other hand, was a fine, dark soil, carpeted with native grass with a scattering of grey rocks. Pecky's wanderings had always been limited to the small picnic area. But this time she must have gotten bored and gone off beyond her usual perimeter. We scanned the hillside above us and then walked over a large area of it. She blended so perfectly into such a landscape that it was possible to overlook her even while gazing right at her. We began to call but as she seldom had found it necessary to reply we did not expect too much from that effort. Moreover, her call was weak and ventriloquous. It would not necessarily tell us the direction if we could have heard it.

We looked at each other in dismay. Supposing she had gone into the willows and along the stream, either up or down—it would be utterly impossible for us to crawl among them or to spot anything through the tangled branches and leaves. The water was not deep enough to drown her but she would have had difficulty maintaining her balance. The danger of stepping on her in crashing through the undergrowth was equally great. There was only one very probable clue—she had to be on the ground as she could not fly, and any elevation would be a low hop onto a willow branch. She could not be high in a pine tree.

We decided that Laney as the smallest of our trio would try to make her way up the stream in spite of the willows. Dale and I fanned out across the other side of the gulch on the dim chance that she could have walked down to the water, crossed, and gone up among the rocks to the east. It was the closest to looking for a needle in a haystack that I have ever come, and I have never used that figure of speech since without a vivid sense of what it really means. Our first fright turned to panic and then became desolation. She simply could not be left out there. True, she had survived once upon a time and perhaps under even more dangerous conditions—in the city—but two miracles were most unlikely.

Dale and I then carefully crisscrossed our way up the east slope, each time going a little higher, never sure whether she might have gone straight up or veered to north or south. Fate was kind and I should have set somewhere a votive candle in gratitude. As we retraced slowly down the rocky slope, in spite of my tears, I caught a movement on the ground. There, but five feet in front of me was a little grey bird on its way up the mountainside, picking its way calmly among the bits of disintegrated granite. Quite unconcerned, as supremely confident of her safety as though she had been going somewhere in the living room at the Lodge, was Pecky. Any fright she had, indeed, would have been caused by my scream of joy and by being so swiftly gathered into my hands. How she had managed to negotiate the stream and more, how we could have been so engrossed that we failed to check on her for such a length of time were mysteries for which we had no answers. Needless to say, she never had another opportunity to provide a solution, for we kept much more constant surveillance ever after on the hill or up the gulch.

Pecky had unwittingly caused us one tremendous fright. Perhaps not so unintentionally she disturbed my equanimity a number of times that winter of isolation in the high country. These moments occurred during periods of quiet companionship between just the two of us when the children had gone to bed downstairs and Pecky and I were alone in the big living room upstairs. After observing that I had settled into the high backed armchair with my reading or sewing, Pecky would walk over to the chair, jump up on my lap, then to the arm, and from there to the back which was level with my head. There after a few moments of "putting up her pincurls" she would settle herself comfortably into a dumpy little ball. She refused to go to sleep, however, and some evenings at close intervals would talk under her breath in little grunts and muffled squeaks. Frequently this sort of talk was a scold. In the deep silence of a winter night her almost whispered comments seemed like a growl and I often wondered what sounds of movement outside she heard and disapproved of which I was unable to detect. Did she feel the need to confide what she knew? What was she telling me?

Part of the deep yearning I have to communicate with other forms of life includes the urge to know and appreciate their store of information and learning. What a breadth of understanding this could create, as well as a more profound conception of this world, not to be restricted to what only our own limited senses, our conditioned lenses tell us. Alas, we cannot even open ourselves to a sympathetic grasp of how other human societies see the world, nor even to those who hear another sort of rhythm in our very own midst. Our severe human limitations only increase the necessity to empathize with other species than our own. In this planet's present precarious circumstances, recognizing and acting upon the urgency of that need—to deepen and broaden our comprehension—could mean the very survival of life.

Though Pecky was an "educated and well-traveled" bird (she even learned to bank as we went around curves in an automobile!), physical prowess and the usual beauty of her species were denied her. One of the loveliest sights in the bird world is the streamlined flight of the mourning dove, the long, tapering tail fanned, showing its black and white stripes, the wings in wide and curving spread giving glimpses of the exquisite blue and pearl-grey feathers on their underside, and the small head far out on the slim neck.

But streamlined poor Pecky never could be after the accident which brought her to us. Her tail feathers seemed to be forever getting broken off, leaving her looking somewhat truncated. Her wings of course never managed their graceful birthright, since the broken one was incapable of spreading to its full length, and her whole posture due to her more or less quiescent life became somewhat dumpy. But her eyes were something else, as beautiful and perfect as any wild dove ever possessed. They were very large and round, sparkling bright, such a deep chocolate brown they seemed black unless seen in very bright light, and were rimmed with a tiny beading of delicate grey. Her eyes had two covers, the blue-grey membrane so thin it barely hid the dark iris, and the heavier sleeping, grey lid. Those huge eyes in the small head could and did reflect pleasure, anger, humor, mischief, and affection. Intelligence, comprehension, an inner dignity shone in their depths. She was probably not beautiful to anyone but us.

We ourselves were responsible for some temporary defects in Pecky's physical appearance. As she lost her fear of us she no longer felt it necessary to flee when we came near nor even to get out of our path. That meant we had always to be on the alert since she expected us to know that she was right under foot. She sat quietly on the couch or in her favorite chair and trusted that we would not sit on her.

Unfortunately, we almost did just that, though not too many times. On several occasions she lacked a tail or had a bare spot on her back because, as we had almost stepped or sat on her, she scooted out in panic from underneath us, leaving her feathers behind her. We discovered what seemed to be her instinctive capacity to loosen the feathers from the body to make escape possible. We were terribly ashamed of those bare spots, during the two weeks of regrowth "recovery," which reminded us every time we looked at her of our carelessness.

Whether her feathers were in tiptop shape or not, her dignity and self-assurance remained unruffled. A feeling of seniority was no doubt developed during the year after Kerry's death when she was the only non-human member of the family. We watched her assert this priority when Kirk entered the household in the summer of 1953 after we had returned to 2049. An ebullient piece of collie pup, he pranced innocently into what Pecky considered her domain and she promptly took care that he understood his place by puffing out her feathers, hissing and administering a sharp peck on his little black nose. He observed the utmost propriety thereafter.

Pecky was a dignified four-year-old when we introduced her for the first time to Jemima and Mimer, the fuzzy yellow ducklings who had just joined the family. With no realization of their impertinence they ambled over on their out-size "pudders" to meet her. But, her rights of seniority now well entrenched, Pecky puffed to her largest size, purred until she shook, hissing the while, and sharply pecked first one and then the other on their yellow bills. Her quick rebuff taught these wobbly, noisy little intruders to treat Pecky with due humility. The lesson was not forgotten and though Jemima and

Mimer grew to ten times the size of Pecky they were always courteous and a little wary of her. Her territorial imperative was satisfied in an unnatural framework.

The whole phenomenon of Pecky's perfectly normal physical career of egg laying (and its volume amounted to a career), provided us an insight into her natural urges, as well as revealing some conduct which her abnormal living conditions no doubt brought about.

It was not possible to ascertain, as Lang had told us when Pecky was found, if she were male or female. The iridescence which identifies the adult male mourning dove is of a most subtle and sometimes almost indistinguishable intensity over that of the female. Being handled so much by us changed Pecky's shading slightly as the oil from our hands darkened her otherwise blue-grey and grey-brown feathers. We kissed her on her breast so much that lipstick invariably added its tint to her own coloring! Looking back over the years it seems impossible that we could ever have failed to see how completely feminine she was, just as Wigga in the next decade was so uproariously and unmistakably male.

Pecky's first exhibition of sex play came in January of her first year with us. She was seven or eight months old. Its suddenness rather startled me because it so transfigured her. I was lying on the couch after lunch, reading a magazine. Pecky was on my leg as usual when I heard a peculiar croak and looked down to see not our gentle Pecky but a wild bird, whose feathers lay flat along her head as far as her neck, then abruptly erected giving her a square appearance. Her under breast feathers were pulled up tight, revealing the little buff-colored knee pants. Her tail was fanned and she was beating a tattoo on my leg by drumming her feet. Her beak was open part way like a chicken's in hot weather. I actually felt the hair rise on my own head, the change was so astonishing.

Through repeated experiments following this exhibition I found it was necessary for me to be wearing jeans, that by drumming my hand on my leg she could pretty regularly be induced to play in this manner. She would hop over my hand and whirl completely around as she landed, drum with her feet three or four times, acting as though she were going to bite the hand beneath her. After each of these jumps, turns and croaks, she would fluff her feathers out, hesitate as though she expected something else to happen and then after long delay she finally gave a vigorous shake. I remember watching our chickens shake in this manner routinely after the rooster had left the hen.

One day I put my other hand over her as she stood croaking. She squatted exactly like a hen, raised her wings to a level, her tail moved to one side, and the feathers parted, revealing the rhythmical opening and closing of the orifice. It seemed at last, unmistakeable that Pecky must be a female dove. Intermittently over the months following, Pecky went through this same routine, but sometimes widened her activity by a few abortive gestures toward what appeared to be nest making. A couple of sticks or straws, a paper clip, a rubber band she brought to an area she chose, pulled them close around her, and this seemed to satisfy whatever nest-making proclivities she possessed. As doves make notoriously flimsy nests she may not have been too far off the mark. She sometimes settled down on the conglomeration and once in a great while went through the motions of regurgitating her food. One time she industriously pulled paper clips out of their box on my desk and then sat on them with a preoccupied air.

There was one procedure in particular which baffled us when Pecky began her actual career as an egg layer. She developed a weird obsession with piano keys prior to the actual egg laying, jumping onto them and walking back and forth. In the mountains, where the action of the old upright piano was so easy, Pecky made a tune as she walked along. But in Denver my Steinway had considerably harder action and she with her four little ounces never produced a whisper of sound. Later on, her unpredictable choice of places to lay made me fear that she might lay her egg on the keys. It could easily break and we would have a mess on our hands.

With all the spasmodic sexual activity she exhibited it was not until June 26, 1953, that she actually produced her first egg. She was then two years old. The birth of this first egg we had the honor of witnessing. She was sitting that morning on the terryclothed window sill of our breakfast room. She had already gone through her peculiar change of disposition and had also made the usual pathetically inadequate preparations for a nest in several places during the previous ten days, none of which were on that window sill. She puffed out her breast, raised up as straight as she could and put her tail up almost at right angles to her body. She did not seem to mind our feeling the robin size creamy white egg which was moist and almost hot. She then tried to sit on it, using her beak to pull it to her and partially under her ruffed breast. But somehow things were not quite right, she looked puzzled, annoyed with the egg, turned her back and walked to the other end of the window sill. Nor could any coaxing revive a sense of responsibility toward her creation. This was to be her habitual reaction to her eggs. Once laid, she had completed her job!

The following day we went to the hills, and that afternoon about

36 hours after her first egg, Pecky laid her second—just where she happened to be—on the arm of the couch in the small sitting room at the Lodge at Rollinsville. The fact that wherever we went Pecky went also changed her habitat frequently and arbitrarily, and may have accounted for the fact that she laid her eggs in all kinds of places during the rest of her life.

Rid of her second egg, Pecky settled back to her normal patterns, and one might say we heard no more about it. But about ten days later she was right back at it. Her third egg was laid again in the mountains, this time on the keys of the old piano. Almost 48 hours later, her fourth was deposited in a corner of the big living room at the Lodge. It was beginning to be obvious that it was important to keep a close watch on Pecky's whereabouts during those times. Another ten days and she went through the whole cycle for the third time, this one while we were in Denver. She disposed of the fifth egg in a dish that sat on top of the china closet, and her sixth on the kitchen floor behind the door. Pecky had a fourth and final round for that year in October, laying the seventh at the Lodge in the mountains and, furthermore, laying it in Dale's hand. The expected eighth was laid again in Denver with the usual time lapse of between 36 and 48 hours.

We learned later that what had seemed abnormally active attempts at reproduction were not necessarily so. Mourning doves have been known to raise four sets of two youngsters each in a season. But Pecky did not have the full complement of stimuli and her constitution was certainly not that of a normally healthy bird. I don't know how normal the eggs were other than being unfertilized, but we did find almost the entire contents to be yolk and thus very different in proportions from a hen's egg. She began laying eggs some years as early as April and the last ones were sometimes laid in October. Her peak year was 1955, with eleven accounted for. Altogether in five years she laid 37 eggs!

She ate regularly during her entire life with us the crushed egg shell we provided. But one of the symptoms of forthcoming eggs we learned to recognize was her hunger and eagerness for the added calcium, when she emptied her egg shell dish in record time.

Though we fed her everything she ever showed a desire for, I am sure there were elements necessary to her nutrition she could not tell us about, which we therefore did not provide. There may have been, in the spring of 1957, symptoms of malnutrition or other deficiencies which should have warned us, but Pecky seemed to feel and to act as healthy as she had for almost six years. She was no less interested in her surroundings, no less active in her participation in our life together, nor was she abnormally enthusiastic or peppy. She had begun the usual egg-laying cycle and had laid the first of her second round. So we were ill prepared for the results of a misdirected flight which she made from the window sill of the breakfast room, a flight no different than she had made many times previously on her way to her feed box in the kitchen next by.

This time she overshot, hit the door, and landed on the bare floor with a thump. Many times before, indeed, she had bumped into things. Her wing had also bled before as it did this time, at the bony protrusion formed from her first accident. We applied some flour to the bleeding which had been a successful treatment in the past, but this time it did not stop the oozing, which was partially serous in consistency. Alarmed, we foolishly took her to a veterinary who was colossally incapable of helping. He offered the suggestion she was deficient in Vitamin K—and gave us some drops to put in her water.

But Pecky drank almost nothing from then on nor did she eat. For three days she lingered, still bleeding a little, getting weaker and weaker. And from her sleeping box which we put at the window in the living room so that she could get the sunshine, she tried to climb to us every time we came near, believing no doubt that we would help her. I had never but once felt so hopelessly impotent, so aching with pity, and that was when my Kerry had likewise suffered and believed that I could help. There is nothing that tears my heart more than this simple trust in me, when I fail that belief.

We will never know if Pecky's fourth egg broke inside when she had her last mishap. This could have had a fatal effect. We never found the egg. But since she had sometimes failed to lay her second egg, it could have been that this was not the cause but indeed some systemic deficiency that awaited only a minor upset to exact its toll.

We wrapped the stilled little body in a handkerchief of Dale's and put it in a cigar box coffin contributed by Johnny, our sympathizing next-door neighbor, and carried it to the mountains where, on the hill back of the Lodge on which she had spent so many happy hours with us, we dug a little grave and placed her in it.

In a much more real sense Pecky is not in that box, but with us yet. Each mourning dove I feed on my present hillside home is part of Pecky. Their very motions are Pecky. But should I never see another of her species, she would still be with us, for nothing can destroy the spirit, the person that she was.

There were two contrasting philosophies held by our friends who knew Pecky, revealed by their comments about her. The first was, "My, what a lucky bird she is to have found a home like yours." The second, "What a marvelous experience you have had living with her." I do not believe the first attitude could ever have helped her achieve the potentials which she possessed and we enjoyed. Other forms of life are quite aware of and fully respond only to a true sense of fellowship in the human animal. It communicates to them, unerringly, that we accept them as equals in our environment, are part and parcel of them, that we grant without question their right to dignity and to the blossoming of their personalities.

Her adaptation to the various environments in which her human family moved and in each of which she was included was a part of her intelligence, but evidence as well of her complete acceptance of us. We were her world—so wherever we were she was at home, at 2049, in my parents' Denver home, at the Rollinsville Lodge, on the mountainside, and even, in 1955, on a ten-day summer journey to Yellowstone and Teton National Park. If she knew that she was utterly dependent for her very life upon our care and forethought, it did not make her subservient; and this I treasure, for it was proof that our belief in equality of rights became her realization.

But how can one put on paper adequately the pixie dowager she became, the affectionate little wisp that was our Pecky? In her going she left a gaping hole, a vacancy so real as to make, for a long time, the passing from one room to another a torment of loss for us. Here was her favorite chair in the study in the corner of which she was used to sit for hours; the kitchen window sill where she ate her cornmeal and asked for baths; here the corner of the Oriental rug where she spread her feathers to the sun in the mornings—her little habits which had become part of us would never again fill each day.

A semi-invalid she had been, but with pride and dignity bearing her poor health, a quiet participant in all the daily family routine and its travels. A sweetness not to be caught in words had been a part of us for six years, a sweetness undergirded by an amazing stamina, fortitude, and determination which pervaded the intelligence of her active and retentive brain, her capacity to learn, to decide upon a course of action and, most of all, to form loving attachments. She taught us lessons whose significance we are still appreciating over twenty years later, the reciprocal power of faith and trust, the strength of gentleness and patience. Above all was her living example of the integrity of her own powers, to be unassumingly and quite naturally herself—her own person. Her unquestioning inner certainty thus made it possible for her to adapt fully and happily to the otherwise bewildering conditions of our human household.

Kirk

Chapter Nine

KIRK

Our good friend Ida who had been so fond of Kerry, was dismayed that we spent a winter in the mountains bereft of any dog (guarded by no more forbidding protector than a half-incapacitated mourning dove); and she kept urging us to get another. We were slow to respond, partly because we were not agreed upon what breed of dog we wanted, though there never was any doubt about our getting another one eventually.

After we returned from that year in the hills, and were living once again in Denver, Ida took matters into her own hands and arranged for a friend of hers, who raised Scotch collies, to bring over a pup from the most recent litter for us to see. This woman claimed her collies were from AKC stock, that we could get the papers which provided the "blueblood" lineage, but that she did not have them because she had not been able to afford the price of registration. This was a believable enough argument, as I recalled the high cost of Cindy's membership some 25 years before. But as we also did not have the money necessary we made no more effort to obtain the registration papers for Kirk than my husband and I had done for Kerry.

There is no more appealing phase of existence, no matter what the species, than the budding juvenile. This small, woolly, tan and white ball of animation with the slightly pointed nose, rolled around the living room carpet, climbed clumsily into our laps as we all sat on the floor to enjoy him, and immediately clambered into our hearts. His owner and breeder was asking twenty dollars for him. We compromised at $15.00. Dale and Laney offered to take from their meagre little banks five dollars each and if I matched it we could each have an equal share. The lady agreed and departed, leaving us with twelve years to come of devotion, intelligence, tolerance, companionship and fun, as well as periods of deep anxiety, all of which no amount of money could have measured.

Since Kerry had derived his name from the supposed country of his origin, we thought it appropriate to choose a Scottish name for this puppy. (I have been told since that Scotch collies as we know them in the United States are not to be found in Scotland. And I must admit I never saw one there.) We finally agreed on "Kirk." I had once known a young man with that name and decided that even though it was the Scottish word for "church," if a man could be called that there was no reason why a dog should not have it also. "Kirk" had a pert sound which characterized perfectly this alert, playful puppy, yet it had dignity which would be appropriate in later years. So many times an animal is named for a juvenile trait which sounds silly in later years.

I am not sure that somewhere in his lineage Kirk did not have shepherd blood and therefore could not qualify for 100% collie ancestry. His nose never grew to the exaggerated narrow length prescribed by the dog show "experts." And his eyes, though deep-set were larger than the slits permitted in show collies. Further, his ears absolutely refused to turn over at the tip and insisted on standing erect. We had once gone through the ordeal of trying to make recalcitrant collie ears tip over in the prescribed manner. We had put gum, used tape, and all manner of weights suggested by helpful friends on Tammy's ears (my mother's collie) only to see them pop back up as soon as the "corrective" was removed. So we never even attempted such nonsense with Kirk. Besides, his ears were beautifully edged with dark fur that gave them character and complemented artistically the deep brown, distinct widow's peak between his eyes.

With each passing year I believe more in the importance of good blood lines in all forms of life, whether plant or animal. There is a healthy pride, an inherent dignity that goes along with this inheritance, observable in all species. Nature has her own methods of ensuring good blood lines. They are sound and demonstrably successful. It is man who has perverted, for short-range purposes of his own, the blood lines of plants and animals he has domesticated. There are various popular breeds of dogs who now suffer congenital, structural or systemic weaknesses and/or tendencies to certain diseases. These perversions have been unwittingly invited because some judge at a dog show has decreed that a certain characteristic of stance or color or breadth of shoulder or leg line is desirable. Immediately, that blue ribbon has signalled dollars to the breeders who can reproduce those externals. With what side effects? What attention has been given to the intangibles of sound and healthy emotional and mental qualities which Nature has been concerned with in her mutations? My mother's collie had a bad liver; our Kirk, it developed, had poor kidneys; my brother's registered collies have had sight and hearing and digestive problems, plus being bundles of highly strung nerves without much common sense.

Kirk was anything but phlegmatic, but neither was he overly nervous. He was an active, energetic, fun-loving puppy who had a distinct sense of game. I almost ruined his enthusiastic approach to life, however, at the very outset of our association. I marked him forever with an ineradicable panic at being left alone inside a building, even though it might be our whole house.

Perhaps, remembering Cindy's tranquil and well-adjusted first night with us, I insisted, with a determination I cannot now defend, that Kirk must start out the same way. We gave him an old coat of Dale's to lie on, shut him in what had been the kitchen of our former downstairs apartment, and expected him to go to sleep. But Kirk reacted at the other end of the emotional spectrum. The poor little fellow was terrified, lonely, and loudly miserable. He howled, he whined, he barked, he pawed at the door, he dirtied the floor repeatedly in his fright and dumped over his water dish. No amount of reprimands or scolding subdued him. His voice was hoarse by morning. He was simply inconsolable. So was Dale, whose bedroom was right across the hall from Kirk's "prison." After a couple of such nights—and mornings after—Dale and Kirk prevailed and the pitiable puppy was released from my Procrustean fixation. He was allowed to sleep by Dale's bed and we heard no more remonstrance.

But marked he was. For we could never leave him in the house alone. We went once to my parents for a dinner across town, leaving Kirk at home. While we were there our next door neighbor called and wanted to know what we had done to the dog. Such a commotion he made that she had heard it in the next house full two lots away! It had been a commotion all right. The windows in Dale's room, being basement ones, were just below the ceiling. Kirk had tried to reach them, and to do so, had jumped on the bed, from there onto a book case which he tipped over, howling unceasingly the while, and the clatter had aroused our elderly neighbor who was noticeably hard of hearing! Thereafter, no matter the weather, we had to leave Kirk in the backyard which, though fenced, seemed not to affect him as an unbearable constraint, since he never tried to escape over it. To provide for rain or snow, we left the yard door of the separate garage at the back of our lot open so he could have shelter. This compromise was agreeable to Kirk. But what a train of psychological difficulties I was responsible for by my stubborness.

The first of Kirk's lessons in co-existence came almost immediately. He was six weeks old, or thereabouts, when he came to live with us; all puppy, a bit lumbering, feet outsized, his black nostrils like a button at the end of a nose that did not give much hint of its future length, floppy ears almost always pricked, and a skinny tail with an impudent white tip, also not indicative of the graceful plume that would characterize him later. His bright, dark eyes spotted all sorts of interesting things to be explored, and one of the first was the quiet, grey-brown bird who lived on the floor at his level. So he pranced unsteadily over to investigate, ears pricked, tail wagging with the uncertain gyrations of a small puppy.

Pecky firmly punctuated her seniority and her territorial rights and became thus to this very young juvenile his first and most firmly implanted "no-no." He was not slow to learn and I do not remember ever having to admonish him again in the remaining years he and Pecky shared. He maintained a distant respect for this small creature. Nobody could have foreseen it at the time, but Kirk's initial lesson stood him in good stead all his life. He seemed forever destined to live among birds: Pecky from the beginnning; then when he was two, Jemima and Mimer, the Pekin ducks who came to live with us; and finally Wigga the pigeon and Gobi the Mynah in 1963, when Kirk was an aging ten.

Since Kerry was already an adult dog by the time Dale and Laney could even play with him, they had never had the training experience of raising a very young puppy. Because they had shared equally in Kirk's purchase I wanted them to feel equal responsibility for his development, so we decided that when Kirk was six months old, he, Dale and Laney should go to Dog Obedience School together.

I am reminded now of the Indian family we visited in Bombay in 1960, who had an impossibly impish, obstreperous, obnoxious child of three. Their two elder children in their teens were delightful, well-mannered, friendly young people, thoughtful, and altogether charming. The contrast was so obvious that the parents felt called upon to explain if not apologize for the pandemonium caused by Previn, whom we privately referred to as Previn-the-Gremlin. They said they believed in allowing him complete freedom until he reached the age of five, whereupon they would begin training. I figured if the system had worked with the two older ones, they must know what they were doing. But it looked hopeless.

Perhaps the same psychology is sound for Obedience Schools, that obedience training does not being until the young dogs have reached a minimum age. At any rate, there were some Previns there, both human and canine, and one can only hope they too benefitted by the training experience before it was too late. Kirk was never a Previn/Gremlin. But he and the children did both need the training of consistency, and repetition, and firmness.

Kirk learned very fast to "sit," "lie," "come," on command. He never added the extra fillips to the commands as did a French poodle, who, instead of letting his leash drag after him when he obeyed the summons to "come," stopped and picked up the leash and carried it proudly in his mouth as he came to his owner. Kirk did as might be expected: well, if not brilliantly. He wanted very much to understand, and to do what he was asked. He learned that those sessions were not for socializing with the other dogs, but for concentrating on what Dale or Laney instructed him to do. Maybe his later serious absorption in getting his charge home from the grocery was at least in part the result of what he learned at school.

The experience was never totally successful, as Kirk never managed to "stay" when we came to that lesson. But this was our fault, because the practice periods which were supposed to be held at home every day between class sessions turned out to be more fruitful ground for arguments betwen Dale and Laney than necessary drill work for Kirk. They either fought over which one was going to work with Kirk, or fought to be excused from it. And I failed to carry through *their* training as I should have.

Kerry or Cindy, either one, would have felt the whole affair an indignity, and Kerry's obedience was so unique anyway, that it would have been quite superfluous and truly an insult to our relationship, his and mine. Several friends who had known Kerry scoffed at the idea of our patronizing Obedience Training for Kirk. Let us say, it didn't do him any harm; I think Kirk would have been just as cooperative without it.

When our two tiny, yellow-fuzz ducklings graduated from their indoor box to the wire enclosure we set up in the backyard, Kirk was enchanted. He spent hours lying outside their pen, with totally absorbed interest, his nose pressed so close to the wire he bent the tip of it, watching the two unsteady babies as they dabbled in their water pan, or ate, or pudded after each other. When they had chasing games he would get up with concern, and hang his long head over the wire, his tail wagging earnestly. They soon became accustomed to his big, shaggy presence, and were unafraid even when they got close enough to the fence that he could nuzzle them.

Later, when they were big enough to be let loose in the yard, he proved as trustworthy with them as Kerry had been with Pecky. He generously shared his backyard, and everything else that belonged to him, with the two ducklings. Jemima very early realized that Kirk was quite a tolerable companion, a protector when necessary, and great fun to play with. Even his exploratory nudges which bowled them off their unsteady legs as babies were evidence of benevolent curiosity which they accepted as such.

After our momentous endeavor of building a pool for the ducks, Kirk unquestioningly felt he had reciprocal water rights. But Jemima and Mimer did not agree, and asserted their territoriality over their pool decisively. When we cleaned out the pool Kirk found it amusing to snatch a few drinks at the edge while the water was newly clean. In fact, it was more fun, he thought, if they were going to make an issue of it. He watched the ducks with mischief in his dark eyes as he lapped the water. Sometimes Jemima lowered her neck in a long straight line, in what Dale called her "destroyer" form, and scooted across the pool after him. It was one-up for Kirk if he got his drink before she managed to reach him. Sometimes Jemima had to get even by climbing out of the pool, waddling after him and nipping at his rear skirts. If he decided to meet the attack head on, she rushed into his full ruff and grabbed, pulling with short sharp tugs. Mimer either stayed in the pool, offering a series of warning suggestions, or if Jemima continued the mock battle too far away, he clambered out and pudded after, at a discreet distance.

Teasing was indeed a turn-about affair. If Kirk could take a drink from their pool, then it was fair enough that they should dog his footsteps, walk on his tail and feet if he were lying down, play tug of war in his long ruff, have a game of tag, or pretend to chew up his playthings or dabble in his water pan. All of this Kirk bore with the good natured tolerance which was so characteristic of him, exhibiting a gentleness that made it unnecessary for us ever to worry that Jemima or Mimer would come to harm.

When she initiated a confrontation, Jemima would lunge at his big white ruff, and nip, retreat, and nip again. Kirk would stand still, pretending to ignore his tormentor, until she got to his skin, then he would turn swiftly and take her long neck in his mouth. But he never closed his jaws, nor twisted—it was all in fun, and a sham battle. Mimer always stood a safe distance away and b'watted his remonstrances or his advice, we never knew which, as Jemima pursued the scuffle.

Kirk could be very sober also. Life was not all just fun, even before he began to have his serious health problems at age nine. One of his most enjoyable sources of relief from the routine of his backyard was the shopping excursion to the grocery store just two blocks away. The sound of his leash being lifted from its accustomed place brought him either from the far corner of the yard or anywhere in the house, his dark-rimmed ears pricked, eyes sparkling, plume wagging. At times we tried lifting it noiselessly, but he could and did hear it. (Kirk lived in the dog leash era and from earliest puppyhood had known no freedom in the city such as Kerry had enjoyed. So he thought it no indignity, as Kerry had viewed it, but quite natural to go out the back gate pulling on the chain.) Mimer always waddled along the walk with him, quacking nervously, and then followed the length of the fence to the corner of our property, there to stand plaintively b'watting his disapproval of Kirk's departure. I had a secret feeling that Kirk enjoyed leaving his uninvited "shadow" behind; it gave him for that little time a special status, which he seldom achieved, for all his life he had to share our attentions. First there was Pecky who had seniority as long as she lived; then came the ducks when he was two; Wigga and Gobi in 1963 when we was an aging ten years. Mimer and Gobi both outlived him.

So the times when he was our sole companion were especially important and significant to Kirk. He played to the limit en route to the store, by pulling us into various byways, including innumerable stops to inform all interested parties that "Kirk has been here," by watering every bush, tree, fence post, fire plug and telephone pole with a seemingly inexhaustible supply of evidence.

We tied him to a doorstop outside the grocery, and always managed to have a package he could carry when we came out. He took it solemnly, adjusted it so he could hold it comfortably, and started off in a sedate walk. It seemed as though he never looked to right or left, never saw a fire plug or a tree trunk, having only one thing on his mind. Mimer always met us at the gate, and hopefully tried to work up a game with Kirk as he walked up the narrow sidewalk through the middle of the yard. He would either cross in front of Kirk to get his attention, in which case Kirk merely walked over him, upsetting him unceremoniously, or dumped him if Mimer tried catching at his ruff. He completely and utterly ignored his erstwhile playmate, to Mimer's unfailing bewilderment.

Kirk pursued his dignified pace up the walk, to the back door, waited to have the door opened, and climbed up the steps to the kitchen, where he deposited with singleness of purpose, his responsibility in front of the kitchen sink, and at my feet if I had remained at home. Mission accomplished, he waited for the praise we always gave him. Sometimes his package held a bone for him, which we then unwrapped and gave him. But whether it had a tempting smell or not, Kirk never offered to get at the contents. His job was delivery, and he fulfilled it 100%. He might drool over its tantalizing odor on the way home and present a moistened package, but it was not molested. If, long back in his genes, there was a working species, as I am sure all dogs were thus employed at the begining of their domestication, then this cherished ritual exemplified those traits in Kirk's character and experience.

As most dogs do, Kirk loved to ride in a car. For him it must have been additionally enjoyable because he was usually the only one of our animal family to go along. On one of Kirk's early jaunts with us in our Ford 2-door sedan, he probably wished he had not been invited. Dale and Laney and he rode together as they always did, in the back seat. This time we had invited guests, two Indonesian VIP's, on a tour of the United States sponsored by the State Department. They were sitting in the front seat with me and we were on our way to the Lodge at Rollinsville. It was February, and raining; but the rain had unexpectedly turned to ice in the first tunnel we came to. As we entered the tunnel we spun, and spun again, crashing against one wall and bouncing back to hit the other side. The door sprang open and our two guests were thrown out on the pavement. When we finally came to a halt, the children and Kirk were in a heap on the floor in the back, frightened and shaking. The two guests, fortunately, suffered only abrasions; but it was a terrifying experience for everyone, including the Ford, which was totalled.

Kirk shook for hours, until friends came to our rescue and delivered us home. Some six weeks later in a new car we essayed our first journey over that road since the wreck, Kirk, the children, and I. Fully a couple of miles before the approach to the tunnel, Kirk, usually silent, composed and alert during a ride, got up and began to whine, and his whole body began to tremble visibly. The children tried to calm him. We were afraid he would try to leap out of the car before we negotiated that tunnel. However, upon passing out of it safely, Kirk settled down again and said no more.

How did he know where we were when he began to show fear? What kind of picture must he have carried in his brain, of the entire route, of the scenes that passed just before that accident? And to put the whole thing together again, to connect it? He must have been able to reconstruct the preceding several miles, then the accident, and pull it out of his memory as one complete picture. His anxiety decreased, but only gradually, as we repeated the trip over the coming months.

Two years after our automobile accident, Jemima left us. Until she died Kirk's interaction with her and Mimer had been only on the edges of their serious and absorbed duck business. But in the weeks when Mimer was coming out of his mourning for Jemima, the little white Pekin began to transfer his need for companionship to Kirk. So far as we could determine, it was neither Kirk's idea nor with his encouragement, but rather because the big, shaggy, brown and white collie did not repulse Mimer's overtures. He was, indeed, both mild and patient. Maybe without knowing it, and because his was a loving disposition, Kirk began to respond positively to Mimer.

Mimer gradually drew close to Kirk and began to initiate and share games that had formerly been Jemima's preserve. He would nudge Kirk's ball around with his bill until Kirk dashed in, grabbed the ball and trotted off, his head held high, mouthing the ball around and around. Kirk would try for our attention by dropping it at our feet, suggesting we throw it to him. It sometimes seemed as though the ball would go on down his throat, he caught it so far back in his mouth. After this game was started Mimer would waddle back and forth between us and the dog, never able to catch up.

In the final round of their rough-house games, Mimer learned to trust the big collie with a completeness that was touching, even to allowing Kirk to take his head or his neck, whichever was handier, in his mouth, an intimacy which would have been unthinkable when Jemima was alive. Their "togetherness" in the backyard, day after day, was a lesson in both tolerance and understanding, and especially admirable on Kirk's part. For it was Kirk who now had to learn to share and do everything with such an unlikely pal as a white duck. Everything that is, but his breakfast. That was Mimer's "no-no," but he always tried his luck anyway, and sometimes succeeded.

In fact, Kirk's breakfast of a half can of dog food was instituted as reassurance that he was not being neglected, and as a reward for having to see Mimer eat his hardboiled egg in front of him every morning. We tried the simultaneous offering also as a practical means of keeping each one contented with his own, but it became a race to get as much as possible of the other fellow's instead.

Kirk was permitted to eat the morsels of egg that were scattered on the ground during Mimer's voracious attack on the egg in our hands. But Mimer apparently felt that if Kirk shared his egg he ought to share Kirk's food. When Kirk's pan was set down, quick as lightning, Mimer would dash in and grab whatever he could manage before Kirk chased him off. Sometimes Mimer had gobbled up such an enormous bite he had trouble swallowing it and looked dazed until the lump got a ways down his throat. Kirk in turn, after bolting his food with one eye cocked on the progress Mimer was making, came as close as we could allow, and stood there drooling and begrudging Mimer's every mouthful until he was permitted to clean it up.

There was competition, but the big collie and the little Pekin developed as well an amusing working partnership in their territority of the backyard, which I am sure had facets and overtones understood between them which we could never even observe or appreciate. They became actually a very efficient team of watch/

duck/dog which functioned against anyone going down the alley, or anyone walking along the front sidewalk, which was more remote but not beyond their mutual comments. The garbage collectors, because of the clanging, discordant clatter they made, merited their most vociferous objections. But the utility meter reader was not far behind, because he had actually to invade their yard. He carried in his little black book the careful notation that at 2049 there lived a big collie and a white duck. Their names were written there, and the comment that they should be spoken to and placated if possible before entering the yard. It worked sometimes. When it did not, and Kirk backed up by Mimer refused to let him in, I had to go out and hold Kirk some distance away. Mimer usually followed us to stand nearby, but he quacked his disapproval as long as Kirk growled. Kirk felt that the mailman and the front doorbell were also within the purview of his authority, and so, automatically, did Mimer.

As Kirk grew older, he allowed Mimer to assume a sort of (Distance Early Warning) DEW-line function. If he deemed Mimer's first quack as a warning worthy of investigation he would look up. If Mimer continued his alarm Kirk would then get up to inquire further, at which signal Mimer proudly increased the tempo of his quacking. Kirk turned over to Mimer more and more of the initial share in this security system. As Mimer had always been more alert to probable duck trouble than Jemima, he thrived under the arrangement.

Aside from dutiful objections more in the line of duty, Kirk had his normal number of personal dislikes that varied in intensity. There were certain individuals, certain things and noises that simply got his ire up. Mimer would pud along after his friend trying his best to share Kirk's indignation. I often wondered if Kirk ever felt the effectiveness of his displeasure was diminished by the participation of this improbable companion, as Mimer's little six pounds was stretched to its highest alongside Kirk's lunging sixty over the back fence, when he threatened the garbage man with mayhem. One of his perennial nuisances he found to be Inky, the black spaniel across our alley who was addicted to incessant barking, and to idiotically racing back and forth along his own fence. Kirk disliked him as thoroughly as ever Kerry had, but Kirk had less chance to work out his feelings because he did not have the freedom to roam which Kerry had enjoyed. So, until I could put a stop to it (since I have never been able to tolerate a barking dog, including my own), Kirk and Mimer would taunt and insult, replying to Inky's pointless behavior across their mutual barriers.

With unvarying consistency Kirk also regarded the helicopters which occasionally roared with their annoying and penetrating put-put right overhead, as unforgiveable invasions of his air space. He believed, apparently, that the old Roman law was still applicable, which granted to a man the rights to his surface property as well as his rights to the earth beneath and the air above. Mimer waddled as well as he could keep up, unquestioningly assisting Kirk in this verbal defense.

Kirk reserved his most violent hatred and implacable intolerance for the power mower that belonged to Johnny, our next door neighbor. Kirk was vey fond of Johnny, and Johnny of him. He was indeed one of the few people outside our family for whom Kirk showed an actual affection. Like Cindy and Kerry before him, he was polite but diffident toward friends and even more so toward strangers. He could have cared less if they petted him or even paid attention to him. But Johnny was different. When he came home Kirk would run to the fence and lean over it to talk to him and to be petted.

But when spring came, and with it lawn mowing time, Kirk forgot all about this friendship. The sound of that power mower just set him wild. We wondered if he even realized it was Johnny pushing it. He became so incensed he almost choked in his own wrath, and snapped and growled and ran along the fence and barked without surcease in as foolish and frantic a pattern as Inky; he would have taken on that mower with his bare teeth had he ever jumped the fence. He could so easily have leaped it that I was always amazed he somehow never did.

Kirk was like a mad dog; he paid no attention when Johnny talked to him; he did not hear me when I called to him. It never failed that I had to go out, literally grab him by the scruff of his neck, and drag him into the house. Even then his anger was slow to subside. He was insulted by my scolding, frustrated by the hated noise he could still hear, and he threw himself dramatically, in very un-collie fashion on the floor, there to growl under his breath as long as the mowing lasted. It was the power mower, not just the mowing process, because he never objected to our old fashioned, manual mower. He did not like it any more than he liked my vacuum cleaner, but neither of those so infuriated him.

Kirk took in stride (i.e., he neither objected to nor particularly embraced) the many guests from around the world, who had begun to frequent our home several years before his entry into the family. Dale and Laney and I had discovered, beginning in 1946, the challenging, stimulating, broadening, and altogether heartwarming experience of building what we came to call our World Family. Men and women from every part of the world became part of our everyday life, and while they spent time studying in this country they also learned how we lived and they and we both learned how very alike we all are.

It became possible in 1960 for us to make a world pilgrimage, to visit in turn these many ramifications of our human family. But before we could consider going for the period we planned, from June of 1960 to July of 1961, we had to make satisfactory provision for the care and happiness of Kirk and Mimer. Ida had early on offered to take Mimer, but as she had two dogs of her own it was not feasible to add Kirk to the ménage.

Two of our World Family members from Norway were at that time attending the University of Denver just a couple of blocks away from 2049. Bernt knew that I had thus far been unsuccessful in finding adequate care for Kirk as well as anyone to stay at our house during our long absence. One day in May, when Bernt and his Norwegian fiancee Astrid were visiting us, Bernt offered in his slow, direct, and deliberate manner, "If you want, I can stay at your house, and take care of Kirk too. I am fond of him and he likes me."

And it was settled forthwith, a happy arrangement for each and every one of us. Not only did Bernt stay, but two weeks after we left, he married Astrid. So we had the best of all possible solutions: two fine young people who loved Kirk and took the most devoted care of him, and who as well kept our house in excellent condition. I think they rated higher marks in both categories than I merited many times in discharge of the same responsibilities.

For one thing, being Norwegian, they loved to walk, and without fail took Kirk for an evening stroll, which delighted him and provided a healthy break from his backyard confinement. They wrote to us at designated stops on our trip, and we became very homesick when we would read about things Kirk did or how he was feeling. One time they invited my mother for dinner, and she reported on Kirk in her next letter to us, which we received on arrrival in Calcutta.

I have often wondered what our Indian host Hemen must have thought of us, sitting there in his office (we were friends of his younger brother and he had never laid eyes on us before), devouring news from home with an eagerness intensified by our absence, by that time amounting to nine months. I was reading my mother's letter aloud, when I came to the part where she described Kirk's reaction to her, his puzzlement and bewildered look because she brought memories that somehow didn't fit with the two people who were now his loved companions. It was as though she were a vision out of a former life that drifted through his head, she wrote, upsetting the furniture of his present existence. He repeatedly came and sat before her, gazing so intently at her, trying to put together his acquaintance with her that did not belong with the two people who now lived in his house. I broke down in the middle, and handed the letter to Laney who also dissolved in tears. Then Dale tried. And finally we apologized to Hemen, both for our rudeness in reading in front of him, and our display of emotion. For the reserved Indian, trained to allow emotional outbursts to occur only in the utmost privacy, it must have been unseemly, even though Hemen had spent some years in the United States. But he was also the perfect gentleman and host and the incident was passed over without comment.

We knew then for certain, when we let down our guard, how homesick we were. One can understand then how we built up our anticipation of the boisterous welcome which Kirk would give us as we stepped onto the lawn at 2049, four months later. Even the neighbors had gathered to watch our homecoming. It is true, Kirk was boisterous, he barked and yipped and wagged his tail, and kissed us, even jumping on us which was usually not permitted. But I felt that there was something missing. It was not somehow as wholehearted as I had expected. I tried to dismiss the impression and to think that I had been expecting too much. He continued to be glad to be with us as we settled in, and certainly we could not doubt from the first moment that he knew and remembered us.

Bernt and Astrid had moved, the day before our arrival, into an apartment near the University. They came over to see us one evening a very few days after our return, and Kirk rushed to meet them. He gave to them the greeting we had expected to get! Then, as he turned from his effusive and whole hearted welcome, he looked at us, then back at them, and again at us. Of a sudden the mists lifted and all became clear in his dog brain. We *were* his family after all, though this did not diminish the affection he had for Bernt and Astrid, the ones who had been his family for the past year. He came galloping over to us and told us unmistakably by repeating his cries of joy that everything was now straightened round in his thinking. Maybe he was apologizing too for his previous lack of full realization. Most of all, I think, Kirk's temporary mixup was a tribute to these two dear young people, that Kirk had learned so to love them they indeed became fused with his original family even when we had come back to him. His confusion was also understandable because he had been left in all the familiarity of his own home surroundings and the only change had been the people. Had he been taken somewhere else and left with Bernt and Astrid, this fusion of personalities and relationships I feel sure would not have occurred.

In the early months of 1962, following our return from the long trip abroad, it became evident that Kirk was beginning to have some health problems. We were reluctant to believe that he was getting old, though this was his ninth year. But we could not ignore the fact that he was having difficulty digesting his food, and seemed frequently to be losing his dinner. He was listless and his fur lost its sheen. By what malevolent fate we were propelled to the particular veterinarian we were, I am not now sure, but I shall never cease to regret that we subjected Kirk to even one hour of his questionable ministrations.

If ever the healing profession, whose degrees should mean the same standard of ethics applied to all forms of life, spawned an unprincipled, coldly calculating, and unfeeling practitioner, it was this man with the deceptively low-key, calm, and deliberate manner who specialzed in diseases of the pets of the rich!

The "doctor" diagnosed Kirk's difficulty as kidney trouble, which meant, he said, that for the rest of his life he would have to have subcutaneous shots of a pint of liquid—water mixed with some concoction or other—every other day, or he would not last a week. However, he was careful to let this blow fall gradually. At first it would be necessary for the next few weeks; then it stretched to the next month or so; then, as we had not yet blown a gasket in indignation, he passed the life sentence.

A major part of my eventual wrath stemmed both out of sympathy for Kirk's suffering and admiration for the stoical endurance with which he faced each trip to that office. All spring, all summer, and into the fall he knew what the preparations meant, and never once in all that time did he make a fuss about going, nor make a scene in the waiting room, nor whimper or struggle when the enormous needle was jabbed into his side. He was utterly miserable when he came home, his sides heavy with the injected fluid and he would lie for hours under the cherry tree, motionless, looking at me with mute misery in his eyes, in every line of his dejected body, uttering not a cry. His depth of understanding, his unwavering belief in us, his determination to cooperate made it even harder to watch him so silently suffering.

But Kirk walked into the veterinarian's office, well knowing what awaited him, almost as though he were sedately going calling. The shots were not all. He was supposed to take a quota of pills and these he also manfully swallowed, without protesting. We would put the pill way back on his tongue and if he did spit it up he would dutifully pick it off the floor and chew it! Such Spartan endurance

I somehow did not expect from him. But adversity brings forth qualities in individuals of all species which may be most surprising and certainly admirable.

After months, and hundreds of dollars of fruitless "doctoring," I was convinced that my growing distrust of the whole setup was more than justified when I overheard one of his helpers say one day, "I feel like a heel, having to do this again," meaning another shot. Well, I spared his conscience, because we never went back.

Shortly after my fortuitous overhearing of the assistant's remark, I complained in my anger and vexation to a friend who happened to call one evening. She immediately suggested her veterinary with whom she had had most pleasant and helpful relations through a number of her animals. So we promptly took Kirk out to the edge of town where this kindly doctor had an office extremely modest in comparison to the luxurious establishment of the former. He gave Kirk a thorough examination, and said he did not consider the treatment he had been through to be warranted by anything he could find, and forthwith stopped it. Kirk lived for three more years and never had another shot.

It was in 1963 that the great upheaval occurred in the lives of every member of our varied family. We decided to forsake our native Denver and to build a home in the mountains, high up on the side of a canyon west of the town of Boulder. It had to have a name, of course, and so we requested a close member of our World Family to suggest one. Soleiman is a Libyan whose Arabic is exceptionally beautiful and whose sense of the fitness of language is that of the true poet, which he is. *Dunyana* means "Our World". To us it was the most significant name we could have chosen, for this home was to represent our personal, intimate family world, as well as embracing what we conceive of as our place in the entire world.

For Kirk, even though an old man of ten years, the move was largely an adventure with unexpected dividends of new scenes and increased companionship with us. The day we left 2049, after almost eighteen years of pain, joy, and growing under that roof, Kirk's chief emotion was the fun of riding with Dale in an old, rented truck, sitting proudly in the cab on the seat beside him. He forgot about his uneasiness, intensified by Wigga's panic, over the strange men invading his house to take away the familiar furniture he had lived with all his life. (Wigga was the pigeon who had burst into the family circle that spring.)

We had had to rent a small cabin in which to live until our new home was completed. Life at the little cabin in the woods proved to be quite to Kirk's liking, providing as it did, a larger area to roam in and many more exciting scents and animals than he had known except in short sojourns at Rollinsville. The cabin's setting on a steep hill also offered Kirk opportunities to get away from Mimer's attentions which had not been possible in Denver. For if Kirk chose to nap too far down the rocky hillside, or too far up behind the cabin, he knew Mimer would not be able to negotiate the bumps and hollows, and he could rest undisturbed.

We could never know how or where Kirk's first confrontation with a skunk took place. We could only know for sure the outcome! Laney had come up from the University with a classmate, a young man of impeccable attire and unflappable manners. Dale had also come from the construction site of our home down the canyon where he was learning about the art of home-building from our Swedish carpenter-contractor. We had just had our dinner and were sitting outside on the tiny porch when we were assailed by an unmistakable odor.

Looking very crestfallen, Kirk came round the corner of the cabin, preceded by a nimbus of piercing aroma. In all the years of living at Rollinsville, among numerous dogs who in their turn had rather pointed adventures with a series of porcupines, I had never been faced with the problem of cleaning off the results of a skunk. Dave (the guest) and Laney filled a bucket of water, and, armed with a bar of soap, bravely took Kirk a little way down the hill and tried to wash him off. He was cooperative but the smell was not. Even Mimer pointedly avoided his adored companion. In the scrubbing process the smell somehow impregnated Dave's clothes and he later confessed that he carried skunk with him all the way back to Denver.

Dale in the meantime called the oldest resident in the canyon for advice, since our remedy had only spread around the difficulty. The earthy advice was—tomato juice, applied liberally. It was then too late in the evening to go to a grocery store and Kirk simply could not come in the house. So he, disconsolately and miserably aware of the odor he could not escape, spent one of the few, if not the only, night of his life shut outside the house when we were in it. The next morning Dale was on hand as the store opened its doors. When he set two large cans of tomato juice on the check stand, the owner of the small grocery looked up at him and grinned, "Hangover, or skunk?"

The tomato juice bath was the final indignity so far as Kirk was concerned, insult piled upon injury. But the remedy, like so many folk medicines, was successful. Thereafter, I kept a stock of tomato juice on hand, which we had to use for that purpose only once afterward.

The skunk episode must have been particularly irritating to Kirk, because he took obvious pride in his appearance, as well he might.

As he grew older he grew more beautiful rather than less, as his earlier golden brown coat became a darker, richer sable each year, making his thick white ruff even more outstanding. He showed this pride in his every graceful movement, his deliberate care in lying down, placing his dainty white feet close together in front of him. Kerry always threw himself in a heap and crossed his front paws in an engaging manner. One was not tempted to tousle Kirk who was a trifle aloof in his dignity and whom one merely petted on the head or stroked gently when he lay down. Yet he needed and asked for love and approval every bit as avidly as Kerry, but in his own more reserved way. He would come quietly to my side, push his nose under my hand for a stroke. He was crushed if I ever scolded, and retreated until he regained his equanimity.

Kirk had never before Dunyana lived under conditions where he had to assert his territorial rights with members of his own kind. Those rights had always been defined, for him, by the fences inside which he lived; and at Rollinsville, while there was no fence, neither were there any other dogs during his lifetime. But he celebrated the day of our move to Dunyana by the most colossal and horrendous territorial battle he ever waged.

Sunshine Canyon, at the top of which we had built Dunyana, harbored and continues to suffer from, far too many unmanaged dogs and cats. At the time we moved in, a group of these dogs had formed a pack which, having nothing better to do, went about in search of mischief. A new dog in the neighborhood offered a spicy challenge, so they all trooped up the drive that led above our house, and their leader bristled down to let Kirk know who was top dog in these parts. Kirk's mettle, as I say, had never been put to the test under such conditions.

We were preoccupied inside the house, and did not see the arrival of the gang. Things happened so fast after that, we had trouble reconstructing the scene. Kirk advanced, stiff-legged to meet the intruder, who at the crucial distance attacked him. All of Kirk's pride turned to fiery indignation. For being completely inexperienced in hand-to-hand combat, so to speak, he turned in a very finished and decisive performance. He early went on the offensive and gradually backed the invader up the hill until he had him at the base of a big pine tree just below the edge of the road. There, while the rest of the hoodlums stood gaping and motionless, Kirk pinned the leader against the tree and drove for his neck. He was a mass of furiously agitated sable and white fur in double-quick time, first lunging for a leg, then the neck, and at all points between, slashing wherever he could close his jaws on the enemy. Savage growling, so loud and ferocious filled the air I could not even recognize my own dog. This was his near-attack on the garbage men or the lawn mower in Denver carried to full fruition. The other dog was yapping, then howling, but Kirk showed no mercy or let up.

We finally intervened, not to save the gang leader, but to spare Kirk any further effort or an injury. He was not a young dog, after all. What would the effete, suave veterinarian who had prophecied invalidism and a quick demise, have said, two years later, to see this ball of fury demolishing an accomplished fighter? We dragged Kirk off his victim. The defeated bully got up dizzily like a prize fighter who has been down for the count, and slunk away without a backward glance. His brave cohorts had already retreated well before him.

We never saw any of that bunch again. Kirk remained in undisputed command over the top of the canyon. But he did carry a scar for the remainder of his life, about an inch long on the left side of his nose, where he had been raked by the other dog's tooth. Kirk had reserves of power and determination with which we had not credited him sufficiently.

His concerns, fortunately, were more often pleasant than not, and sometimes so cooperative they were very funny. Our second winter at Dunyana, after a very severe snow storm, Laney and I were trying to remove the chains from our car out in the garage. We had made a ridiculous mess out of it, and gotten the chains tangled over the axle. Our troubles took us both under the car. When I turned my head, I saw Kirk down on his front legs, peering at us, trying to see what was so wrong that we were crawling and struggling underneath and out of reach. He was so earnest about it, his long head cocked, and his question so plain, that we exploded into laughter. And the mess seemed to sort itself out with much more ease after we sensibly relaxed.

His cooperation extended to a willingness to shake hands which neither Cindy nor Kerry had thought worthwhile. He not only was willing, he purposefully put out his paw when asked to do so, and tried to hook it into your hand, pursuing his effort repeatedly until he connected. He apparently saw nothing undignified in it as the others had.

Nor did he think it beneath his dignity to "bow" on command. As collies frequently stretch by bowing, it was easy to turn this natural act into a consciously peformed one. He had two bows in fact, one a deep salaam, his front feet clear to the elbow flat on the floor, his head close to them, the other a mere curtsey, a quick half dip, which he used if he were in a hurry. If Mimer were on hand, the duck interpreted the salaam as an opportunity to waddle over and playfully nip Kirk's nose, a one-sided pleasure we were sure, but one which Kirk characteristically allowed his friend.

He had taken Wigga's entrance into the family with his usual good nature. He had felt free, when Wigga was on the floor, to nudge him almost off his feet and sniff him rather unceremoniously, while at other times he allowed the impudent pigeon to take liberties with him, such as parading up and down his ribs, as he lay by the door, or letting him do his round-and-round inches from his nose.

Kirk was becoming more absorbed by his own discomfort and related difficulties at this time. He began to have trouble at night with incontinence; I did not at first connect the dark spot I would occasionally find on the old Navajo rug by my bed. But it was finally unmistakable. He could no longer wait until I let him out the door. About that time an old hernia on his belly that we had all but forgotten about from his youth showed up a dark and menacing blue-black bubble. Our kind veterinary, now a considerable journey away, said it had become a "strangulated hernia" and suggested an operation if we wished, but warned that at his age Kirk's chances of coming out of the operation were about 50/50. It might with equal probability, he thought, never reach crisis proportions. So we did nothing.

Kirk's functioning machinery continued to break down. He bore with his marvelous patience the wearing of a diaper at night which I had to institute, and he chose to endure without comment what must have been an almost insufferable embarrassment to him. I of course never commented either, but praised him extravagantly if we woke in the morning to find him dry. There is no point in detailing the anxious weeks and months of his ups and downs. As with any failing loved one in the home, our days were centered on his care. Any plans we made were with his state of health that day the primary pre-condition. He grew weaker and weaker, less able to climb even into the car, and he began to avoid going up or down the stairs. He lay outside the kitchen door most of the day if it were warm. Mimer hovered close by and would pud over frequently to speak softly into Kirk's face, as though he were asking, "How is it today, friend?"

At last one evening he stood weaving unsteadily in the kitchen, seemed not to hear us as we tried to coax him to his bed in my room. So we lifted him, Laney and I together, and carried him in, where, during the night, he passed away. I shall never forget the leaden weight that pulled at me when I saw that still form which would never move again, nor look at me.

Kirk's ashes rest in an urn beside those of Kerry.

I wrote to Dale who was in Korea. "They slip away so quietly, these loved ones, and all of a sudden there is the empty shelf where we used to look up and see Wigga almost always, and the blank sandstone outside the kitchen door with no more brown furry body lying there. . . . Poor little Mimer, he stayed in his pool for two whole days and would not leave it. He frequently, now, looks around inquiringly as though asking where his pal has gone, and then he quacks, low and soft." With his highly developed animal intuition, I am sure Mimer did know Kirk had gone as had Jemima before him. Just as the wildlife on the hill knew, somehow.

Within 24 hours, the foxes and raccoons came close in the early evening, the chipmunks gathered, the rabbits. We realized then that when Kirk used to lie outside the kitchen door at night, peering intently with his ears pricked for all the sounds he could gather, and occasionally lifting his long nose to sniff the air, these animals were just beyond the pale of our back flood light. They were regular visitors and diners from that time on.

Sometimes the ache of a loss can be sharper when a quiet, gentle, and dignified character whose tolerance and patience seem to make no calls upon our time or attention slips away than when a demanding individual dies. One wonders then, "Did I do or give all I should have?" And so the sorrow after death is, as always, a form of self-pity or self-reproach. And each loss can fortify our determination to live, enjoy, give, while we are blessed with that presence.

The immediate wave of wildlife which flowed over us showed me that a part of that unique value of living at the crest of a mountain for which I had indeed come would be utterly sacrificed by taking on the raising of another pup. Kirk had been too old to chase wildlife away, but his very presence had kept any from coming near.

So a new phase—a physically dogless phase—of my life began when Kirk left us. I am not dogless in the larger sense, for my dogs *are* with me, and always will be.

Kirk's immortality of influence is also wider than my own heart and memory. I wrote Bernt and Astrid in May of 1965 to tell them "our" beloved Kirk had left us. Astrid replied in her Christmas letter from Oslo that she and Bernt ". . . had never somehow been able to forget a certain sable and white collie whom we grew to love so much. So when we decided to get a dog for our family, we tried to find one that was as much like dear Kirk as possible." Thus Kirk's memory lives on in Chantie, a beautiful, delicately proportioned brown and white female collie who does indeed look so like Kirk that I cried a little when I met her in Oslo in 1974. Kirk lives in Chantie, and Chantie will live in the memories of Morten and Stina, Bernt's and Astrid's two children.

Leibnitz

Chapter Ten

LEIBNITZ

While at college Dale had, during 1959, a roommate who was fascinated by lizards and chameleons with whom he generously populated their dormitory quarters. Whether the general reptilian world began to interest Dale, or whether he felt he needed a pet of his own in self-defense, I am not sure. He purchased a small green turtle about an inch across and just a little longer, on whom he fastened the unlikely and somewhat ponderous name of Leibnitz, mainly because he was studying at the moment the famous philosopher's contribution to mathematics.

Before long, Dale gave up the competition with the lizards, and Leibnitz made his way home to us. He was living in a shallow glass bowl about eight inches or more in diameter with a rim perhaps an inch and a half high. He had a rock in the middle on which he could climb to get out of the water, but otherwise life was pretty monotonous and lonely. There was a can of turtle food, whatever that consisted of, and he was supposed to like lettuce and raw meat, hamburger suggested.

I felt like the lame helping the halt when Dale returned to college, leaving Laney and me with a dependent little creature about whose needs and habits I was totally ignorant. Leibnitz handled his new situation with what must have been good, survival-oriented, common, turtle sense. He retreated under his mobile-home shelter and poked his head out only for food, or to take a swim and break the boredom by climbing up on his rock to sit.

He grew rather imperceptibly, but we realized after a few weeks that he had grown, because we found him jerkily creeping along the table top, having somehow negotiated a climb over the edge of his bowl. He was, however, so tiny yet that we had to check frequently to make sure he was still in his bowl, as he would have been most difficult to find had he fallen off the table and gone wandering.

There seemed almost no other avenue than that of food (and the act of feeding) by which we could reach out to this tiny turtle with the dark, mottled pattern on his shell, and his light-buff, soft underbelly with its spotted, "finger-print" identification. He had the merest excuse for a tail which appeared from under the shell, and it along with all other moving parts could retract under the umbrella almost instantly on alarm. His claws were surprisingly delicate for being attached to ungainly legs; legs that had to operate (at such an awkward angle) from under the handicap of that hard carapace. His head was about the size of a large pea, flat on top, but with slight ridges at the sides. His eyes were a source of wonderment to me. They seemed to see so much, and they were such microscopic pieces of optical technology. How infinitesimally small and precision-perfect were the parts that operated them—another example of Nature's awesome attention to the exquisite pattern and functioning of the smallest unit!

Leibnitz had only one area of his body where there seemed any possibility for interaction with us, and that was his head, which could retreat so far within that he looked headless, and at other times could gradually emerge at the end of a long skinny neck to weave back and forth. So, when he did send his head out to its full distance from his shell, and especially when we offered a morsel of hamburger during his feeding, I took the occasion to touch the top of his head with a finger, very lightly. The first time he drew it back with more speed than I had seen in him yet. If slow motion is the secret to long life it is no wonder turtles have enviable longevity.

In a moment or so he cautiously poked it out a little to assess the situation. Since nothing happened and the enormous human who stood above him offered no harm, he decided it was worth hazarding for the meat he very much wanted. My second attempt he chose to ignore and opened his jaws, razor rimmed along the edge of his horny beak, and very quickly snapped up the tiny bite I held out to him. He was, on the whole, rather careful not to bite us when we fed him and became more so with time. I graduated from a light touch to a light stroke and we progressed in a very few days to deliberate communication.

I have never been able to toss food to our animals and go off and leave them. Feeding is one of the most strategic times to get acquainted, to observe, to teach, to become more intimate. To make feeding a time of pleasant ritual is the quickest and easiest method I have found by which to cement the ties of trust and affection. But it has to be done sincerely, not mercenarily, and for the intrinsic reward of its own moment; i.e., the human must *enjoy* giving pleasure (the food) to the other animal. The animal then quickly comes to associate a glad human contact with the satisfaction of eating. Animals are uncanny in their recognition of sincerity and honesty; they can sniff a dishonest, ulterior motive with almost unfailing accuracy, and they react accordingly. Food must nourish the soul as well as the body.

Leibnitz discovered that he liked the feel of stroking; nothing dangerous appeared to happen when he was enjoying this novelty; and soon, he was actually inviting the sensation. I would lift him out of his bowl, all his moveable parts hidden entirely, hold him unprotesting in my hand, and stroke his shell. Whereupon his little head would emerge, he would look at me sideways with those incredibly small, pinhead, observant eyes, and promptly extend his neck just about as far as it would go—and we were friends. Dale claimed I "de-turtled" Leibnitz because turtles just don't do things like that. But I like to believe that I pro-humaned him, or destroyed the fear that was a natural part of his defenses.

He learned to know when we came into the room. In fact, I think he very likely looked for us, he had so little diversion otherwise. When he saw or heard us he tried to get our attention by swimming mightily around, making waves in the water, bumping into the sides, stopping to stand on his hind legs at the edge, quite obviously searching for a way out. Sometimes he clambered up the rock in the middle and looked around. Either way, he gained his point. We noticed and lifted him out for the stroking of his little head which he had come to desire. Perhaps this exciting goal was what led him, as he grew larger, to stumble upon the means of falling out of the bowl in order to search for this satisfaction.

We found another gratification for him at the Lodge in Rollinsville, when we made one of our weekend treks there a few weeks after Leibnitz came to live with us. We decided to house him while there in a larger pyrex baking dish in the kitchen, a much more commodious living arrangement than he had in Denver. We gathered a few, different-sized rocks for variation in his furniture, and after travelling in the cramped quarters of a box for a couple of hours he was more than pleased to acquaint himself with his new and larger accommodations. He swam about, and poked his head among the rocks, struggled up one of the larger ones and looked about. Another advantage of this container for us was that the sides were too high for him to climb out as yet.

We decided we would have to purchase another baking dish for his Denver housing, as he was by this time becoming too adept at escaping his smaller one. But we had not yet done so, when one day we were preoccupied with other activities and neglected to check on Leibnitz whose bowl sat on the table in the breakfast room. He must have gotten lonely and restless enough to succeed in hoisting himself over the edge, to crawl along the table top and fall off the edge to the floor.

From there we never knew where he went. We never saw him again. We searched under every piece of furniture on the floor of that house, including the stove and the refrigerator in the kitchen—which we laboriously shoved away from the walls. We lifted up all rugs he might have crawled under. The thought of his literally starving to death haunts me still. Dale did point out the fact that the back door was open and the screen never did fit tight, that Leibnitz might just possibly have been drawn to explore the outside world, and gone his own way. But we were denied an enrichment never to be offered again.

Leibnitz is a shadowy little figure among the rest of our congregation, for he was with us such a short time; he was also my only venture near his species. He is there, however, gently outlined, because he was so quickly responsive to our overtures. He understood love from one species to another, on turtle terms, and within little, green, turtle limitations. He comprehended, just as all our other animals have, in his own way.

Turtles have been around and managed to survive intact quite a bit longer than human beings—they have been a recognizable species for some two hundred million years. It seems not impossible that in the larger scheme of things, some higher form of evolving life may, on some distant day, look upon *Homo sapiens* as having been even more limited than the turtle. Should man succeed in what is now the very probable destruction of himself and everything else on this fragile planet, we will indeed have proved ourselves to have been much more tragically circumscribed than was a little green turtle.

Chapter Eleven

JEMIMA AND MIMER

Beatrix Potter's engaging Jemima Puddle Duck was one of my childhood's most loved characters. I enjoyed the story no less as a grown woman when I read *The Tale of Jemima Puddle Duck* to Dale and Laney as small children, and we all agreed it might be fun some day to have a Jemima in our own household.

In the early spring of 1955, when Dale and Laney were 15 and 13 respectively, I went away on a short trip. During my absence, the children went to a supermarket with a friend and were fascinated by a tray of tiny Easter ducklings for sale. They decided to buy one as a surprise for me. But it cheeped so interminably and mournfully they went back next day and bought another, whereupon the cheeping turned into contented cluckings, and the two downy yellow balls cuddled close for companionship and warmth.

I was indeed surprised, but as the children well knew, there was no question of keeping them. The gift having been accepted, there was immediately the important choice of a name, and not a moment's doubt that one had to be called Jemima Puddle Duck. It did not matter that the original English Jemima was the Aylesbury breed of duck, and these little ones were Imperial White Pekin. Since they seemed carbon copies of each other at that point, it was natural to choose a twin-type name. It was thus we christened Jemimer, to go with Jemima. But the shortened nickname, Mimer, was easier to use. He was Mimer to us and to hundreds of people thereafter throughout his long life.

Their first small grocery box shelter was quickly outgrown by Jemima whose already adventurous disposition led her to try climbing over the side. We replaced it by a packing box with higher walls and continued to house them for the first weeks in our already crowded study-den-library at 2049.

As the spring days warmed and their ridiculously stubby wings began to show miniature white feathers replacing the original yellow down, we built in the middle of the backyard a chicken wire enclosure like the one we had constructed for Pecky. This one was a yard or so in diameter and a foot or so high. During sunny days when they could be outdoors it was a relief for everyone, since the amount

Jemima & Mimer

of newspaper and the number of changes required in the packing box, due to their duck plumbing facilities, made life inside increasingly difficult.

They had a pie tin to drink from, but alas, they stepped into it, tripped over it, and tried to swim in it more than they drank from it. They grew so rapidly and required so much more space, the chicken wire enclosure was soon removed altogether, and from then on they shared the entire, fenced backyard with Kirk.

While Jemima and Mimer were still very small, we filled the two deep wash tubs in the utility room to let them swim, one to a tub. Their delight was so evident and their reluctance to get out so plain, as they paddled furiously trying to avoid our hands, that we decided to present them with one of the popular plastic wading pools for children.

This might have been their permanent pool, except for Jemima. Mimer waddled up to the inflated edge, pushed his tummy over the top and sort of toppled in. But Jemima, who had already achieved the one step required from our yard into the utility room, regarded the rim of the pool as another step to be negotiated. Finding it somewhat soft and unsteady, she stood on it teetering back and forth. To keep her balance required gripping, and to grip the smooth plastic, she had to dig her toes in. Her sharp nails bit into the material, none too stout to begin with, and out came the air. The rim collapsed, and out flowed the water. The two ducklings ran after the contents of their pool, watching it sink into the ground, and then paddled disconsolately over the crinkled remains, cheeping loudly. As soon as we patched up one hole, inflated the rim again and filled the pool with more water, Jemima would play her teeter game, and we had to go through the process all over.

They so obviously loved their water that we had to provide something more durable for a swimming pool. In October of 1955, when Jemima and Mimer were six months old, and almost full-grown, we decided to invest in the construction of a permanent concrete pool. Dale dug the hole at the far end of our yard, in front of the garage which stood at the back of our lot. One friend donated the cement, another carpenter friend poured the concrete. It was supposed to be two feet deep and an oval shape, but because Dale got tired, it developed into an unexpected kidney shape, which was really rather charming. It also had gently sloping sides at the two far ends, which Jemima and Mimer used to scramble in and out.

Jemima led the alarm over the invasion of her backyard by Milo, the carpenter, the digging which upset her grass, the sacks of cement that appeared and the piles of sand that were dumped on the ground. But after a temporary retreat to the far end of the yard, she saw that Kirk was safely in the middle of everything, and waddled close enough to watch.

It was the bringing of the hose to the pool, after the concrete had dried properly, that sparked her first intimation of what was to come. The hose had meant water filling the plastic pool, and as it gurgled into the bottom of the new pool, Jemima's quacking encouraged Mimer to venture near, his neck held high in what had become his characteristic posture of apprehension. Jemima could not wait; she slid in before the pool was full and began to splash immediately, with Mimer following seconds later. This was their very first wholly satisfactory bath: dipping head first, shuffling of wings, some preening, then more ducking, throwing water over their backs with that amazing ability to carry water on convex heads and necks, and deliver it onto their broad white backs. Then came time for fun and games: tearing round and round the pool, diving, and swimming under the entire length to emerge at the other end; turning in split-second speed to launch a sham attack, all the time splashing as much as possible and keeping up an incessant quacking of sheer exultation. Jemima and Mimer were instantaneously and without previous practice, all grace and expertise personified, now that they were in a large pool with water deep enough they did not have to touch bottom. Further, they had their very own, exclusive territory which they recognized as such and protected from that first filling of the pool.

The two of them stayed in the water for hours that glorious first day. When tiring from play, they quietly floated with heads tucked behind their wings, dozing, the only movement of the water caused by their own breathing, or a lazy push with one foot to propel them toward the center of the pool. There would be an occasional muffled, contented cluck, coming from inside their insulation, and one bright, black eye would open in a field of white, and then close, at perfect peace with the world. Ever after, each year, with the first filling of the pool in spring, there was that same anticipatory joy as the water gushed in and the same rite of celebration was repeated. The $22.50 which the cement pool had cost, though expensive in terms of our small financial resources, was the most spectacularly successful investment I ever made.

As we watched Jemima and Mimer grow, we frequently wondered about their lineage. The official breed name, White Imperial Pekin duck, suggested that there must have been royalty somewhere along the line. My later study did confirm that the White Imperial Pekin came straight out of China, arriving on the east coast of the United States about 1870. The Pekin and the English White Aylesbury (the original Jemima Puddle Duck), as well as the Rouen duck on the Continent of Europe, are all descendants of domesticated mallards. The Aylesbury and the Pekin have come from a white sport, or mutation, of the mallard. Central American Maya history, insofar as it has been researched, indicates they also had domesticated a kind of white mallard, a most interesting fact, inasmuch as the domestication of the Imperial White Pekin by the Chinese is far more ancient, and also appears to be quite separate from that of either the Europeans or the American Indians.

Since Jemima and Mimer were our first experience in living with ducks, their growth and development were a source of daily revelations, of wonder and amusement.

No awkward, gangling human adolescent whose parts seem not to match, ever looked more ludicrous or pathetic than did Mimer and Jemima as they lost their baby fuzz and grew into the skeletal frames that preceded their adult feathers. The rather lumpy figures were severely exposed: protruding craws and pendulous rears, and the long, skinny neck, which always seemed longer on Mimer than on Jemima. Mimer could be identified in every picture we ever took—even though they looked like identical twins to outsiders—because his head was always held inches higher than Jemima's, and his expression was questioning, half-worried. He seemed to achieve this appearance by pulling in the feathers about his mouth and on top of his head, while Jemima's cheeks were almost always fluffed out in contentment.

A certain mobility of features is possessed by every animal I have known. By the movement of muscles around their eyes, as well as by moving the eyes themselves which altered the position of the small, surrounding feathers, Mimer and Jemima could change their appearance to register contentment, eagerness, fright, anger or concentration.

After graduation from the indoor baby-box to the backyard, there was the problem of housing at night. We lived in the city, and while the ducks were in no danger from coyotes or foxes, there were dogs as well as human predators, perhaps more dangerous, that made safety at night imperative. We settled on the garage, a two-car, generously proportioned affair that housed only our single car plus numerous outwashings of a household with growing children. Jemima and Mimer were alloted a near corner of that garage, which we enclosed with packing boxes; their water dish was put in one corner of the "pen," and a matting of newspapers and sometimes straw put down for their bed. Living with ducks or other birds involves dealing with what appears to be their chronic case of diarrhea and makes the problem of cleanliness a constant chore. They were natural, but their setting was not—hence the issue was ours, not theirs.

They liked the garage as sleeping quarters so long as their corner was hemmed in, because the box-wall provided the protecting limit they needed. At dark they were quite ready to go in though they wanted the formalities of being put to bed. In the mornings they were so eager to get out and into the pool or to start on their waddling peregrinations around the yard that we frequently found their bedroom shoved all out of kilter. Jemima's exploratory tendencies would have gotten the better of her, and before we got out to the garage she would have wandered under the car. Since she usually managed to come out decorated with grease, it necessitated our scrubbing her back feathers very gently with soap. This was detrimental to the feathers because it destroyed her natural oil.

By late autumn and cold weather we changed bedrooms and brought the young ducks into the house, where they occupied a corner of the utility room, one which was truly utilitarian in all respects. It was the back entrance to our house, and one low step above ground. Jemima and later Mimer, found it no problem at all to mount that step coming in, though they tumbled going out, and landed on their broad tummies. In fact, Jemima had long ago mastered that one step so easily that she next essayed the five additional ones that led up to the kitchen. One evening as I was preparing dinner, I heard an alarmed and insistent quacking in the utility room. Turning, I found Jemima underfoot, pleased and important as she could be. It was Mimer at the bottom of the steps who was, as usual, protesting.

After her initial success Jemima found it great fun to mount those steps in no time, partly walking, partly hopping, to explore what was going on in the kitchen. Since Mimer was not about to be left below and alone, he learned to negotiate the steps also. But he flew up each step more than he walked, and with considerably more commotion. This accomplishment became almost a nightly exercise, and they were so expert at it that friends began to remark about their self-taught "trick."

This rather innocent accomplishment led naturally but unexpectedly to their great adventure into the theatrical limelight. The P.T.A. at the new Junior High School which Dale and Laney were attending was in need of extra funds. So it was decided in the spring of 1956 that the parents would put on a talent show to earn some money. A number of clever acts emerged, and at the suggestion of some fellow P.T.A. members we were persuaded to show off our ducks and their "trick." Pecky was asked to participate but she was so small, and her accomplishments geared to parlor-sized audiences, that we declined on her behalf.

Jemima had already shown a cooperative if not enthusiastic tolerance for being dressed in a Puddle Duck outfit à la her namesake, and she would even pose obligingly for pictures. Her costume consisted of Laney's red Camp Fire kerchief which went over Jemima's back and tied under her big white tummy. The blue bonnet had belonged to a long abandoned doll, and its strings were tied directly underneath her bill. It is almost superfluous to add that Mimer would have nothing to do with such wardrobe nonsense, but immediately shook off the bonnet, fell over the shawl, and scuttled for the safety of his pool.

We planned to have Jemima and Mimer make a dramatic entrance from the rear of the auditorium, to waddle separately down the two aisles, then climb the steps to the stage which was connected with the audience area by means of a dozen steps, stretching the full width of the stage. It is easy to see how the idea of the "duck act" was born.

To parade two ducks the full length of an auditorium had obviously to be a controlled sort of progress. So we borrowed a couple of Kirk's leashes, and I made two ruffles out of dark green terry cloth, which I fitted around their fat little breasts. However, there was absolutely no way to keep the whole contraption from coming off over their long slim necks and heads. And neither Jemima nor Mimer took at all to the idea of being led. In fact they refused quite effectively by just backing out of the whole harness. So we compromised—and they led us! This kept the ruffles on because they were pulling them further back instead of off over their heads as they strained to get down the aisle. I suspect Mimer was convinced he was running away from that thing that kept pressing around his middle, whereas Jemima rather seemed to enjoy the walk.

We had a dress rehearsal at school to familiarize them, and hopefully, to elicit more cooperation the night of the show. And we set up a practice routine in the backyard. As was to be expected, Jemima complied with relish; Mimer fled to the far corner of the yard and then resisted the whole procedure. He persisted in regarding the leash and the ruffle more in the nature of a noose, but since Jemima was obviously enjoying herself he eventually went along, with however many misgivings, taking his cue reluctantly from her enthusiastic and vigorous lead.

The night of the performance, the two ducks not surprisingly stole the show and became famous instantaneously. Jemima revealed a latent "ham" talent which amazed even us. She waddled down her aisle, quacking proudly as she pulled Laney along behind on the leash; down the other aisle pudded Mimer towing me. He also was talking, but out of a deep and disturbing unhappiness with the whole situation. All he wanted was to rejoin Jemima. What she enjoyed was the people laughing and watching her. They fortunately could not tell the difference and thought both of them excellent performers. Jemima clambered up the steps to the stage like a veteran and pudded comfortably to the middle where her Beatrix Potter Puddle Duck costume had been placed for her. Dale held a microphone to her bill and Jemima obligingly talked into it like a professional. Meanwhile Laney was removing her ruffle and putting on the scarf and bonnet. When the audience clapped, Jemima, glorying in the fact that she was the center of admiration, waddled to the edge of the stage and looked over her fans, flicked her broad tail, figuratively took her curtain call, and pooped!

During the stage performance poor Mimer was only dazzled by the lights and confused by all the people and noise. I held him to keep him from bolting the scene entirely, and could feel his heart pounding, his breath coming hard and fast. His bill was wide open, hot to touch, his tongue held prominently in the middle, as he did when unbearably excited. He lifted his wings slightly to cool off. He could hardly wait to get backstage to the quiet of the dressing room assigned to our performers. But Jemima had no intention of leaving all the fun and cutting short her stardom; she lingered at the edge of the stage until Dale had finally to carry her off. Paper towelling which we had brought along was used to "clean up our act" before the next number came on stage.

Perhaps it was because of their marked differences in disposition which complemented rather than opposed, that Jemima and Mimer were so congenial. They never had to compete; they seemed to us sometimes to be two individuals who shared the same breath and led a tandem existence. Jemima very early assumed leadership in playing with Kirk or with us. But side by side, they hunted for grubs and worms, they preened simultaneously, they swam in a water ballet *pas de deux*, napped at almost the same suggestion, sitting down together in what Dale described as their "non-shock-absorber-plunk," two white blobs of feathers side by side.

The various plantings in and around the perimeter of the yard, and any other place where the soil had been disturbed, were legal hunting ground for Jemima and Mimer, which they grubbed with great intensity, dexterity, and thoroughness. There were at least two complete inventories made per day, starting at one corner or another, and going all the way around, accompanied by much soft talk between them. Their daily patrols kept our yard free of many pests, by perfect biological symbiosis.

We had six currant bushes, whose fruit when ripe was regarded as particularly and exclusively duck property. Jemima was more than indignant when we judged the time had come to make current jelly, and with a pan and scissors began to clip. The bushes were already neatly cleaned of every berry as high as they could reach, which was a good half of the way, but we still got nipped for picking what remained. It took two of us to do the job, because while one picked, the other had to hold the pan and fend off Jemima. One time we momentarily left the pan on the ground, and Jemima had scooped up the contents before we could get to her. She unhesitatingly took us on as trespassers, scolding us clear to the back door, trying to reach the pan which held what she regarded as rightfully her property and not ours, loyally backed up in her indignation by Mimer—from a distance.

As we watched and enjoyed Jemima and Mimer mature, they seemed, in their perfect harmony, almost like two sides of a single life. However, it was unmistakable that, unlike Jemima, Mimer was a very non-contact person. He was always quite content to let us pick her up and love her, because he could then stay unmolested out of reach. He did not seem to want or need the show of our affection which she joyfully absorbed.

There were a few other differences, not immediately apparent to the casual observer, which told us that they were indeed two individuals. Mimer's feathers were always shinier and smoother than Jemima's, his feet a little trimmer (if indeed a duck's "pudders" can ever be considered trim): that is, the scales on his legs were smoother, his toe nails a little stronger, the webbing a little more orange. Jemima had what might have been called a faint birthmark on her bill, a dark streak in the middle of that overhanging tip which served as such an effective nipper. There may have been other physical differences to which we should have been sensitive, but were not.

In 1957, when they were two years old, Jemima fell ill. It was sadly unnoticed by us because all ducks seem to be afflicted with a constant case of diarrhea. She went downhill very fast, refusing food and drooping only a day or so. Our second and most tragic mistake was in taking her to a veterinary who did not know a thing about birds. Jemima passed away three days later. Diagnosis given: uncontrollable diarrhea or enteritis. But I am not convinced that this was her trouble.

What was uncontrollable and inconsolable was Mimer's poor heart and memory. Jemima's going was to be the first of several traumatic changes and losses in his life. Mimer's adjustment to these events, at what emotional cost we could only guess, nevertheless did bring forth hitherto unsuspected aspects of his strong and sturdy character which had to have been there potentially all along. As is true in humans, neither the ache of his loss nor his keen memory of her were lessened by his development in the years that followed. As Emily Dickinson wrote, "Time is a test of trouble, But not a remedy." So long as we lived at 2049, for six years after Jemima's death, he gave graphic evidence of a continued association with his once inseparable companion.

Every time there was a happening, a filling of the pool, approach of a stranger, a rainstorm, a game with Kirk, a tidbit from the kitchen, he ran, in his funny pigeon-toed side-waddle, to the windows of the basement bedroom which were on his level. There by the back door he could see his whole reflection and he shared faithfully each affair with it. This began only, and simultaneously, with Jemima's absence. His excited quacking would end by his settling down beside the window, and occasionally clucking to it. When the window was open he became very disturbed because he lost that reflection which was proof to him that his beloved Jemima was there, though unreachable, and somehow so strangely remote.

One day, without realizing what an electrifying result it would have on him, we held Mimer up to the mirror that hung on the wall in the breakfast room. We wondered if this clearer reflection would generate the same reaction as the window. His instantaneous, joyous excitement, attempting to leap from our arms straight through the mirror, to greet his cherished companion whom he could at last so clearly see once more, jumped his temperature to an alarming degree. His bill became very hot, he panted and showed such emotional stress that it was unbearably pathetic. It took some moments to calm him down after we contritely carried him away from the mirror.

In one way, it seemed as though he never quite accepted Jemima's death. She was never absolutely gone, so long as he could see his reflection in window or mirror. On the other hand, with that peculiar sensitivity which so many animals and some humans possess, he may have known exactly when she had left the life they knew together. The unnatural reflections only confused the finality.

He gradually adjusted to swimming, sleeping, hunting, eating by himself, all dependent now on his own initiative. And whether he wanted it or not, Mimer had now to tolerate the affection we had formerly lavished on the willing Jemima. We simply could not resist sweeping him up in our arms and loving him, whereupon he felt it was necessary either to take a bath in the pool or do a dry preen in order to take off the human scent and to smooth his feathers which he deemed we had disarranged.

Nevertheless, he did discover a specific time in the day when he could not only tolerate but take the initiative in asking for the reassurance of loving, bodily contact. Every morning when we knelt down to get him from his bed, no matter where we lived subsequently, he vociferously said good morning with neck stretched, his eyes snapping. This greeting over, he thrust his head as far as he could into our lap, occasionally muttering while hidden, a prolonged, muffled "qu-a-ack." The regulations which he had established next allowed us to pet him, stroking his neck, his back, his wings. He determined strictly the length of time this was permitted, standing absolutely motionless until whatever comfort or encouragement or communion he was needing was satisfied. Then he withdrew his head, and it was time for him to go outdoors without delay, as eager for the day's activity as he had been to postpone it moments before.

At nighttime he reversed himself. We put him in his bed or pen and withdrew in a hurry, because the rules then dictated that his bed had suddenly become his territory; we were within critical distance and he whirled to chase us off, nipping in the process. Territory and critical distance were forgotten, or ignored, or forgiven, in the morning, however, when he wanted and needed to be loved.

The terrible ache Jemima's death left in our hearts was somewhat dulled only because within a few weeks of her death we were totally preoccupied with Pecky's terminal illness. By the time we had recovered some equilibrium, Mimer had begun to develop with Kirk his working and living relationship, which, with the exception of one fifteen-month period, was maintained, deepened and strengthened for the next eight years.

It must be admitted in all fairness to Kirk, that togetherness was more Mimer's idea than his. It was simply, for Mimer, transferring to Kirk the companionship he had always known with Jemima. But Kirk did tolerate the lonely little duck, even though he must have felt his basic rights sometimes impertinently intruded upon; and he appeared to tolerate this intrusion graciously, which was even more commendable. Reciprocally, Mimer did offer diversion which surely must have relieved some of the big dog's boredom, spending his days as he did in a limited backyard, with less than adequate space in which to run.

We felt, however, that for Mimer to be allowed to make mud pies in Kirk's water dish was equivalent to our being asked to share our toothbrush. So Kirk learned to drink his water at a new elevation from the top of an old orange crate, at a height Mimer could not reach, though we often saw him try. His idea for dirt and water was to mix the two equally, and then to blurble happily in the result, getting his bill black and his white face spotted with dark brown freckles.

Not even naps were left out of Mimer's attempt to transplant his habits with Jemima onto Kirk, and he developed a technique by which he achieved this goal. Like our other dogs who had known assorted chickens and cats, etc., Kirk felt it was somehow beneath him to be seen sleeping, voluntarily, right next to a duck. Mimer could not be expected to appreciate that angle, so from the time he lost Jemima, he was determined to substitute Kirk for a napping partner. He quickly learned by experience, however, that one did not go about this too directly. If Kirk were asleep, or getting ready to be, Mimer plopped down nearby, but not too close. In a few minutes, he belly-crawled a bit closer, and kept this up, inch by inch, until he was almost touching him, but not quite. Sometimes the ruse didn't work, and Kirk would get up, resignedly walk a little further, and flop down, audibly groaning. Occasionally, if he had been successful, Mimer went into a sound sleep, his head completely inside his wing, his white body next to Kirk's dark sable coat. If Kirk heard something that needed his attention, he got up, unceremoniously pushing Mimer in the process. Mimer got up likewise, choosing to ignore the discourtesy, whether planned or otherwise, waddling as fast as his short legs would take him, to do his share in the guarding of the yard. If Mimer got in the way, Kirk merely walked on him or bumped into him, upsetting his never too secure footing.

As is the case with all companionships, there are serene days, there are days when we gird to battle the world together, and even some when we battle each other.

One of those blue Mondays, when nothing seems to go right, began with the late arrival of the milkman. He usually delivered so early that neither dog nor duck was yet outside to object or interfere. Dale and Laney and I were having breakfast when we were disturbed by loud, angry barking, accompanied by an insistent quacking in the backyard. Then we heard the milk box door shut. When we got to the back door the besieged milkman was almost on the run to the back gate, with Mimer and Kirk noisily accompanying him on either side. As he closed the back gate he spotted us, and with a wave of his empty milk carrier, signalled his relief at getting out of the way of his two persecutors.

Hardly had we managed to quiet them down when Alarm Number Two, the garbage truck, turned into our alley. Not enjoying the best of mechanical health, the old truck rumbled along wheezily, stopping next door to pick up the garbage with the usual clatter. Kirk had already gone to the attack, and was hanging half over the back fence. Mimer was at his heels. It did seem as though that morning the men made more crashing confusion than usual, as they threw the lid on the ground, and banged the battered garbage can on the side of the truck. With no less noise they slammed it back on the ash pit, all to the accompaniment of the furious, throaty barking of a very angry collie, and the loudest and fastest quacks of which an outraged White Pekin duck was capable. I am sure Mimer could not appreciate just how preposterous he looked, trying to equal Kirk in both stature and vehemence.

I don't know why the meter reader had to come that morning, too. But he did. I had to go out and literally protect him this time, because both animals were in a surly, uncompromisingly unpleasant mood by that time.

And that was the day for a helicopter happening, too. After Mimer had assisted Kirk in scaring it away with louder than usual clamor, he went to the window and told Jemima about it. Maybe he gave her an inventory of all the other naughty things which had piled up to disturb him that day, for it was a long harangue before the basement window. By noon it semed we surely would be hearing complaints from our tolerant neighbors, but they kept their peace, even if they may have felt like suggesting it might be pleasanter if we all moved away.

Having worked himself into a veritable stew with all the annoyances, Mimer was in fit condition to have hallucinations even over things native to his yard. Somehow, we had left the hose trailing into his pool, after we had filled it the day before. We almost never forgot to take it out, but there it was. And Mimer almost never reacted to the hose either. But that day it must have become all of a sudden, a big snake. He raised an enormous fuss over it, and refused to go near his pool, skirting it with great to-do. I had hoped at least that swimming in a clean pool would calm his "shattered" nerves. So when I discovered this new source of distress, I quickly pulled it out; but he took his time about getting in, scouting warily around the pool, grumbling the while.

For a few hours, life settled down to comparative serenity. However, Johnny, our patient and understanding neighbor next door, was off work that afternoon, and his lawn needed cutting. He brought out his power mower, and Kirk tore holes in the flower beds as he snapped and yapped through the fence. Mimer seconded him in high decibel bedlam. I had reached my limit, even if the pets had not, so I summarily swept Mimer up under one arm and hauled Kirk by the neck with the other and marched indoors. Mimer sulked in the utility room, venting his anger by dumping over his seed dish and tipping his water pan, so that the water ran all over the floor.

As Mimer's relationship with Kirk deepened, he came through the wall of shyness because of his need to relate, reluctant as he may have been to do so. He even learned to box with Kirk as well as Jemima ever did.

The jousting usually began by Mimer's waddling up to him, quacking loudly, in a very peremptory manner. If Kirk showed no reaction, Mimer assured himself of getting one by nipping him gently on his cheek, or if Kirk were lying down, on his neck. If Kirk were standing there were no preliminaries; Mimer waddled straight for his ruff, immediately beneath that long nose. After a few tugs, during which Mimer managed to reach his skin, Kirk would growl, tail wagging, and it was then his move. He would turn his head and come at Mimer from the side so that he could get his skinny neck in his mouth. At this point Mimer almost seemed to present him a hold by thrusting his neck forward rigidly. In the ensuing scuffle Kirk managed to step on Mimer's foot, sometimes on both feet. It was hard not to do so, since they were so big. If Kirk stepped on both feet it set Mimer on his tail. If only one, then Mimer pulled with all his might to get it free, squawking loudly. Sometimes he did limp for a while afterward.

The quirk in Kirk's active tail was of great interest to Mimer. After observing it, he hooked his long neck over it in what came to be a form of greeting. Occasionally, instead of observing the human ceremony of shaking hands, as Mimer passed by Kirk, he seemed purposely to be saying "Hi!" by laying his head in the curve of his tail.

How few if any places there are in a city any more where children can have other animals to grow up with, to learn to care for them, to play with and observe them, and above all to respect them as individuals. Mimer was a constant source of learning for us. One who was so essentially shy, and hesitant to try things new, survived with dignity intact a devastating loss, plus later changes of habitat and new companions. This was an accomplishment which not all humans can claim.

It was difficult to be blasé around Mimer, for even in his aloneness he greeted each day with such enthusiasm and eagerness. His first tumble down the step to the outdoors was an adventure, even if it were snowing and he turned right around and came back inside.

The freshness of his approach to each moment of living—the reality of the present—was a lesson not to be forgotten. It carried him through his losses, his changes of home. Perhaps that is what Voltaire had in mind when he wrote, "I envy the beasts two things—their ignorance of evil to come and their ignorance of what is said about them."

Small but important routines constituted Mimer's life until 1960, when for the next three years things, one after the other, were turned upside down for him. His underpinnings of security were removed to a degree that might have felled an individual whose ancestral genes were less hardy than those he carried.

The first upheaval was due to our planned trip around the world in 1960, when Mimer was five years old. The care we arranged for him had to be not only adequate but happy, as far as we could possibly arrange it for him.

Our friend Ida who had figured in the lives of both Kerry and Kirk, was willing and even eager to have Mimer come and stay with her, to be part of the flock of miscellaneous ducks and chickens on her suburban acreage. She had in fact been urging us to "replace" Jemima ever since Mimer had been left alone, and now took on the offices of a "marriage broker," suggesting that one of her own White Pekins might be a suitable mate for Mimer.

It seemed desirable at least that they get acquainted in his own backyard before we took Mimer out to her home. So, one day, we put a big cardboard box in the car and went out to get Mimer's intended "bride." "Ducky" turned out to be, in our eyes, a not very prepossessing female. Her feathers were not so clean nor neat as Mimer's. (Ida had no pool like ours.) She was indeed almost frowsy, and her feet were downright unattractive, having long, skinny, crooked nails. Her bill was not as golden orange, nor as straight and perfect as Mimer's. It was long, almost a dish-shape, instead of having the proper convex shape. Her eyes were not the bold, direct, dark blue that both Jemima and Mimer had. And to top it off, she had a limp. Even worse was her uncouth, hoarse "gronk" that never failed to get a surprised prick of his ears and sometimes a sniff from Kirk.

Mimer may have been a "loner," and lonely, for the three years following Jemima's death; but for whatever reasons, Ducky's appearance or her personality, or Mimer's loyalty to Jemima, or the character he had developed since her death, Mimer rejected Ducky from the first step she took into his backyard; rejected her firmly, quietly, and with finality. He never chased her away, nor bit her, nor was in any way rough or ungentlemanly in his rejection. He simply ignored her. And as any female knows, this is far harder than having something to fight against.

We felt sorry for Ducky, for she seemed to us so anxious to please, as she faithfully limped after Mimer at a respectful distance, wherever he went. If she joined him to eat, Mimer left; if she slid into the pool to swim with him, he climbed out the other side. If she plopped down too close to Mimer to nap, he got up and moved farther away, just as Kirk did to him. Even Kirk caught the mood, but was intrigued nevertheless, trying frequently to nuzzle Ducky and sniff her, but she would go squawking off. Mimer regarded that as a matter entirely between Kirk and her, made no move to interfere, and certainly none to protect her. After three weeks we decided to return the spurned Ducky, and let the two of them meet again on her ground, though we suspected the results would be the same. They were.

Just before our departure, we entertained at a backyard picnic a few close American friends and those of our World Family who were still in Denver, some of whom we would travel with, and others whom we would visit on our world pilgrimage. Picnics, which brought strangers into his territory, who moreover wanted to pet him, were not Mimer's favorite occasion. Since he was such a non-contact person, it upset him and produced a return of his old shyness. He either retreated behind bushes, or to the middle of the pool (his safest refuge), and usually would have nothing to do with the crowd of chattering, alien humans.

I had assured those of our guests who with some apprehension regarded Mimer as potentially aggressive, that he never came near strangers. And Mimer did take to his pool. Everyone had been served dinner, all were sitting on the lawn eating, when some quirk of fancy prompted Mimer to climb deliberately out of his pool. He waddled straight toward one of the *men* no less (men, whom he disliked and distrusted much more than women), shook the water off himself onto those nearest, and promptly and unmistakably demanded some of the man's hot dog. With no shyness evident whatsoever, Mimer looked him straight in the eye and quacked his clearest "please may I have some" message. Fortunately, George was a friend of whom we were especially fond because he *did* understand our pets and liked them. To the great amusement of all the guests who were watching Mimer's performance, and somewhat to our embarrassment, he fed that white duck not only the bread from his hot dog, but also the meat!

Mimer then waddled over to the guest sitting next to George and repeated his request all over again. Jemima could not have been more at home in the crowd than was shy Mimer. His appetite satisfied, he mingled like a gracious and veteran host, and put the finishing touch on his performance by staging one of his wrestling matches with Kirk. This remarkable emergence from his cocoon of shyness, which articulated an express and increased desire to participate in our activities, made leaving him so difficult we almost cancelled our plans and turned in our tickets. Mimer's rejection of Ducky indicated how very alone he was going to be, no matter how kind Ida was or how much love and attention she gave him. What would a year's separation, which from his standpoint must seem equal to desertion, do to this development?

Without much conversation, Laney and I made ready to take Mimer out to Ida. We put him in the big cardboard carton and set it on my lap. As Laney drove we tried to divert our thoughts by being amused at the curious attention he received from passing motorists, or from people on the street if we were stopped by a traffic light. For Mimer did not believe in sitting down in the car.

Ida was waiting for us. We carried Mimer around to her backyard which was amply inhabited by a couple of dogs, an assortment of chickens and ducks, including a huge and rather fierce looking Muscovy duck. The only water was in an old, galvanized tin bathtub, near enough to a size of Mimer's old plastic pool that we were not surprised when Ida later told us he quietly appropriated the tub, took it as a matter of course that he would bathe in it, and taught the rest of the duck colony to do the same. It had apparently not occurred to them before.

We set him on the ground and stepped back to watch things take their course. The other ducks, including Ducky, paid little attention to him, except for the Muscovy, who, we learned, continued to be the one who gave Mimer trouble throughout his stay. Mimer met the population with equal indifference. Looking back at us frequently, he seemed plainly more interested in what we were doing and where we were likely to go.

In retrospect I seriously fear that we were like the over-anxious mother, leaving her child at school for the first day. Mimer was basically more capable of handling the situation than we. Moreover, we could not have been more fortunate in his relocation. Our acquaintance with Ida had come about in the first place because of our mutual love for animals, and our friendship deepened as we discovered how alike our philosophies were. It was comforting, therefore, to know that Mimer would be in congenial surroundings, would be shown the love and the kind of care we would give.

So far as we could detect, Mimer gave little if any recognition to his previous acquaintance with Ducky as he warily moved among these new surroundings, with an eye on us. At last we could linger no longer. We turned to go and moved toward the front of the house and our car. Mimer immediately left the other ducks, and waddled up the hill toward us. Then he stopped, stood still in the middle of the yard, his neck stretched, quacking, not his demanding quack, but asking. The action was so plain, we could hardly bear to say, "No, Mimer, you can't go with us. You stay there." He came no farther, just stood there, on the green lawn, irresolute, seeming to us so very small, so very alone, so clean and white and so much a part of us—we turned quickly and looked no more.

I have always wondered if Mimer knew how hard it was for us to leave him. Or if his limited perception of what was to come merely permitted him to live each day taking things as they came, but holding in that excellent memory his own home and family. Ida's reports, as we pursued our way farther and farther from home, indicated that he did not mourn, that he carried his self-assurance with him, which had developed so remarkably in the three years since Jemima's going. This in itself gave him immediate status with her ducks and chickens.

Mimer remained aloof, took as his due the bathtub water. He ate the hardboiled egg each morning from Ida's hand, as she faithfully continued our habit. He also accepted without fuss the isolation at night from the other ducks, which the Muscovy imposed on him, sleeping alone with the chickens at the far end of the poultry shed. I wondered what his feelings about this discrimination were? Resentment? Resignation? Lonely, or relieved to be alone? So passed the next fifteen months of Mimer's life.

As for us, during our pilgrimage around the world, wherever we encountered ducks—little panhandling mallards in Bavaria, who followed us on a boat ride in the canyon below Berchtesgaden; in Bali, where we admired the intelligence of the little rice gleaners; not to mention Mimer's numerous cousins, the Aylesburys at the home of Beatrix Potter in the Lake Country of England—each encounter sent a pang of homesickness through us. We hungered for reports about him which inevitably seemed too infrequent and incomplete.

After our return we could hardly wait for the day to retrieve Mimer. We drove the same car, carried the old cardboard box lined with newspapers, and appeared around the same corner from which we had disappeared fifteen months before.

He was down at the bottom of the acreage with the other ducks, and registered interest but not excitement when we called his name. He merely stood, while Laney went toward him, arms outstretched, still calling. He allowed her without remonstrance to pick him up, and when she squeezed him too hard he "b'watted" a familiar protest.

We had often said Mimer was philosophical, but perhaps Santayana had the better phrase: "animal faith." He seemed to be neither frightened, surprised, nor upset when we got into the car. He rode as if we had done it yesterday, head up almost level with mine, watching the passing scene. When we pulled into the garage, took him out of his box and set him on the floor, he walked unhesitatingly out of the door. There he turned without delay, dove into the pool, swam across it, and clambered out on the other side. There stood Kirk. Mimer went right up to him, quacking vociferously, pulled on his ruff, and then straightaway, started on his rounds of the bushes and plants along the perimeter of the yard. He stopped at the windows by the back door on the way around, to report at length to Jemima. So he had not forgotten her, either. For good measure he made the journey around the yard twice that afternoon.

That night he went to bed as if there had never been an interruption, ducking under the mangle chair, and quacking little nothings to himself as he settled down in his "hole." Endowed with the marvelous memory which I have observed in all the animals I have known, Mimer could return to the same page where he had left off and reopen it, with not a moment's time out for readjustment.

Kirk spent about ten days in the hospital in 1962, due to his kidney trouble. It was the first separation Mimer had suffered since his reunion with us after our long trip. Perhaps Mimer knew before, better than we did, that Kirk was ill. And who can know what panic he may have felt when all of a sudden Kirk was not there. Just as all of a sudden Jemima had not been there, and he had never again been able to summon her farther than the shadowy, half-real vision in the window. That Mimer was very much aware of Kirk's absence we knew, because after he had gone to the hospital, Mimer sat on guard alternating between Kirk's own ball and his other plaything, the old socks. Mimer would look up at us when we came out to keep him company in the backyard, and would give a mournful, inquiring, low quack, then put his head back under his wing. He almost seemed ill too.

Great was Mimer's rejoicing when Kirk returned to the backyard. He spoke to him at length, quacking directly into his ear, yet he was remarkably careful not to bite his ruff or tease him. Instead he hovered like a protective guardian near him and over him as the big dog lay quietly on the grass.

In the spring of 1963 Mimer was confronted by a new presence with which he had to come to terms. Wigga, a baby pigeon rescued from the railroad tracks by Johnny, our next door neighbor, came to live with us. He grew to adult size in a matter of a few weeks, and thereupon adopted Mimer as Mimer had Kirk. Mimer ignored him for the most part, even as he had Ducky, but Wigga chose to interpret his indifference as acceptance. So far as Mimer was concerned, Wigga was an unmitigated nuisance.

If Mimer ate, Wigga flew to his perch where we kept his seeds and ate also. He would settle on the ground at nap times, as near Mimer as possible. But it was when Mimer was in his pool that Wigga became upset and paced noisily up and down the edge. Depending on Mimer's mood, he either ignored the eager pigeon or came sailing full steam straight for him, in which case Wigga had no trouble getting the message, and would fly straight up, then hover overhead scolding him. He sometimes landed on the edge of the garage roof where he could overlook Mimer in his pool right beneath, from which vantage point he would lecture, in reproof or frustration.

Long before we actually moved from 2049 to the mountains west of Boulder, in October of 1963, Mimer was perturbed by the days we spent cleaning out the garage, that repository of everything we hadn't known what else to do with for years. Packing boxes were set in the backyard and filled. These were not only an intrusion on Mimer's territory, they portended something strange and unusual to his keenly sensitive disposition, even though he could not have known what they meant. About the only stable thing in his life at that period was Kirk.

Mimer's distress the actual day of moving was pathetic. Wigga's panic was perhaps even more keen. Our most recent acquisition, a Mynah named Gobi squawked and flopped around in her cage adding to the general confusion. Strange men were going in and out of the house, all the doors were wide open, Kirk was upset and barking and very seldom in the backyard. It was a very traumatic day for all the pets. As for us, Mimer was the least of our pet problems because even open doors and a gate left ajar did not tempt him to wander. He retreated to his pool as the safest way of avoiding everything. Kirk would ooze out the gate, or wander through the house, and poor Wigga flew in and out constantly, over people's heads, lighting on furniture, gronking as though he could stop all this commotion and put everything back as he knew it.

After the movers had taken everything that went into storage, Dale finally arrived in the dilapidated old truck he had rented from some obscure source in Boulder, to take the few pieces we were going to live with until our new home was finished. Kirk was to ride in the cab of the truck with Dale, to his great pleasure. Laney and I had packed the Buick to the gills, with the incredible assortment of odds and ends that always remain in such cases, and had hitched Tuffy, our ancient jeep, onto the back of the Buick. The neighbors gathered in amused curiosity, to see how we were going to get all the pets stowed away in the remaining space for the two-hour jaunt to the cabin which I had rented far up the canyon beyond our new home. They were kind enough to say that they would miss us and the "entertainment" we and our pets had provided through the 18 years we lived there at 2049.

We lowered Mimer, his bill open and already panting from the day's excitement and turmoil, into the usual big cardboard box heavily floored with newspaper, and I carried him on my lap. Did this operation bring back memories of the trip to Ida's when his family disappeared for so long? Whatever was going on in that rounded, highly arched little white head with the tiny feathers standing almost erect in his agitation, he managed to work himself into a fever pitch over it.

We have often wished we had a record in moving pictures of us all as we pulled out from the curb, late in the afternoon of that October day; left the house that held most of the memories of Dale and Laney growing up, of a succession of pets, and of our years spent in the gathering of our World Family. But no motion picture could have caught all the sounds—the rickety truck ahead, the Wigga gronkings from Pecky's old cage stowed on the bottom of the backseat floor, the hoarse and breathless Mimer quacks, up front, Gobi's rasping protests and occasional screams out of a tiny box into which she was packed between us, and the queer jerking noises of Tuffy's impudent shape pulling on us from behind. Even less could it have recorded the emotions which all of us—furred, feathered, and bare of skin—were at that moment experiencing for our separate reasons.

This ride was the first during which Mimer became car sick. We finally put him on the floor where we thought he would be less dizzy; but settle down he absolutely refused to do and only churned up himself and the papers under him. That was one of the longest two hours I can remember. We could drive no more than 35 miles an hour, not only becaue of Tuffy's weight dragging on the Buick, but because of the asthmatic condition of Dale's rented truck ahead. When it came to climbing the seven long miles and the three thousand feet from the mouth of the canyon to the small cabin where we were to live for the next months, we had grave doubts that we would even get there. By the time we arrived, it was dark, cold, and all the pets were hungry, bewildered, cross, and thirsty. And the cabin was so small that we literally tumbled in, a jumbled mass, and continued to live there in layers.

Mimer had again to make the most difficult adjustment of all the pets. He endured with fair equanimity three months of stacked existence, during which he had neither a bedroom of his own, nor level ground outside he could roam on, nor a pool to bathe in, nor surcease from Wigga's company, nor the comfort of Kirk who was either out of his reach on the hill or with me at the new house.

We left the stacked arrangement at our miniature shelter in the dead of winter before our home was completed, and moved into what we could of Dunyana. We had to transplant Mimer and Wigga to the confinement of Dale's as yet unfinished room. Workmen still swarmed everywhere else in the partially finished house and occasionally had to invade Mimer's quarters, arousing his great indignation. Considering Mimer's disposition, his age, now almost nine years, more strange surroundings, the confusion and noise, and above all being incarcerated with that pigeon again, it was in the nature of a triumph that he came through the ordeal still pretty well hinged together.

During the long winter months at Dunyana, Mimer spent a majority of his days in the kitchen, where we set up his corner by the glass door that opened onto the hillside. As I was also in the kitchen much of the time, I began to realize the range of Mimer's conversational ability, how remarkably complete was his language.

Language comes in many forms, especially if we mean by "language" communication with others, the phenomenon of memory, and the act of decision making which all involve concepts of some sort. We learned from both Jemima and Mimer that not necessarily is a quack is a quack is a quack. It can say numerous things, by length, by inflection, by volume. Benjamin Whorf, the early linguist, once said that no language could be considered primitive, in the sense that it did not meet the necessities for communication adequately, because all languages *have* met the needs of the people or other animals using them at the time and in the circumstances they were using them.

It was because Mimer communicated with us directly so much more after Jemima left that we began to appreciate more fully this instrument he used to manage his life. His vocabulary was undoubtedly different from what it would have been had he lived his life in a natural state. He had to have, for example, a clear mental picture of bread, of his egg, of his bed at night, none of which would have been similar to or included in his mental furnishings had he lived his life as a wild duck.

As long ago as their first year at 2049 we had been treated to a distinct and humorous serving of Jemima's verbal observations. I was scolding Dale and Laney with considerable feeling over some misdeed which had severely provoked me. We were in the utility room where Jemima and Mimer were standing also, quite motionless, listening and watching intently. As I stopped to get my breath, Jemima offered a very long, profound "Qua-a-ck," soft, low, but insistent. Her comment was so aptly timed, its tone so appropriate, that she transformed the tension into humor, and somehow the lecture ceased to be important. (Which is probably why I cannot remember what it was about.)

Their tonal system, like that of the country from which Jemima and Mimer had originally come, was the key to their communication. If one is limited to "quack" in all its possibile inflections, the tones become all important!

Spending a good part of the remainder of his life in the kitchen at Dunyana, Mimer became intimately acquainted with all its operations, just as Gobi did three feet above and nearby him. He had a computerized memory of what went into the wall refrigerator, just above his corner. So long as he knew blueberries (one of his favorite treats) were still there, he would come and stand beside me at the opened door with his long neck stretched to its highest, quacking his request. Sometimes we did not understand each other, and at such times he became visibly impatient and frustrated.

To me his most endearing use of language was the quiet conversational habit he developed. Ever aware of my actions, he came to feel that he should reply not only to a question or remark directed toward him, but even to a glance. His remarks varied from a muffled quack, much as we would say "m-m," to a definite word or even a sentence as he construed them.

All in all, we asked a good deal of Mimer—to be happy, contented and reliable on a rugged hillside which, even if we loved it dearly, he could not be expected to like as he had his level, grass-covered, fenced-in yard in Denver. The least we could do to give him some pleasure was to provide another pool, which we did as soon as we were well enough acquainted with our hill to know where to put it, and how to engineer it. Mimer did have Plastic Pool No. 2 for the interim, which sat in a sad and wrinkled form on as nearly level an area as we could place it. He showed a proper disdain for its shortcomings as he waded about in it, biting occasionally at the rim which refused to stay inflated, even without Jemima digging holes into it. He used it only for business purposes and never sat in it just for sheer pleasure.

We dedicated Cement Pool No. 2 with due ceremony by making Mimer's footprint at the edge and carving the date, July 5, 1964. So successfully did we press his foot into the soft concrete in spite of his objections that his footprint shows the articulated joints of his toes. Though Mimer had shown indifference and even aversion to all the commotion entailed in the construction of the pool, he recognized its significance when we pulled the hose over and began running the water into it. He quacked excitedly, waddled back and forth in his impatience to get in, and immediately asserted firm ownership after his first slide into the water, most especially against Wigga.

Mimer celebrated the occasion with the ritual he and Jemima had shared on the first spring filling of their pool at 2049. It began with a peculiar high-pitched "gronk," which we never heard under any other circumstances, accompanied by a high arching of his neck, a simultaneous rising upright in the water, and treading in place with his feet. Then he launched joyously into the first satisfactory bath he had had for some nine months!

Thorough "soaping" was followed by an in-water preen. A few moments later he clambered out the gently sloping west side, which ever after was his exit from the pool. The toilette continued, by his "ratchet" action, pulling his bill from his flat bottom clear to his neck, the rough teeth along the underside of the upper bill catching all loose feathers and dislodging them. With an incredible "card shuffling" of his wings, plus a hearty shake, no guilty loose feather remained. Application of his cosmetic and waterproofing oil came next, by rubbing the bottom of his bill over the oil sac just in front of his tail bone, then spreading it over his feathers. Next came return to the water for a rinse, and perhaps a repetition of the whole process.

We had brought his old rubber pool ball with us, with which he and Jemima had had so many games, and which he had learned in his solitary amusement to dribble around the edge of the pool, propelling it by the water he blew out the sides of his bill. He greeted it like an old friend and immediately reinstituted his former game. Each week he spent hours at this pastime.

Kirk found the hillside much more convenient to make himself unavailable to Mimer. So he frequently chose to nap, or to survey the canyon undisturbed, some fifty feet away up the hill from the house, in the shade of the old sand wagon we had brought down from Rollinsville (a sentimental journey inasmuch as the wagon had been used to build the dam at the Ice Plant fifty years earlier). To get there required a tremendous effort for Mimer, who therefore only occasionally made the trek. Coming back was worse, because his brakes were even less adapted to downhill walking than his legs to uphill climbing.

As Kirk's age grew increasingly heavy for him to bear, he sat on the mat outside the kitchen door longer and more frequently. Mimer altered his activities accordingly. If he saw that Kirk was preparing to lie on the mat, he waddled quickly to sit on a corner of it, in which case Kirk might decide not to sit on it at all. Or if he were already lying on it, Mimer tried to get on too, with the result that Kirk got up in disgust, and lay on the hill a few feet away, leaving Mimer disappointedly in sole possession.

In the spring of 1965, the kind and understanding vet who had freed us from the malevolence of the earlier one, told us there was not much more time left us to live with Kirk. Did Mimer somehow know this? We found him so often talking gently to Kirk. He did not pull on his ruff anymore to entice him into a game he could not play. If Mimer were in the kitchen, and Kirk lying just outside the door, stretched along the base, Mimer plunked himself as close to the other side as he could, with occasional soft quacks through the glass.

We wondered how Mimer would endure the loss of a second great life attachment, when the inevitable parting took place. Did he know, the night Kirk died, in the room over Mimer's bed, that he would not see his companion again? The next morning he was very quiet when, after his mushing time, we carried him upstairs and set him down outside the kitchen door. He pudded over to his pool, slid into the water and quietly paddled around in it. Nor could we coax him to leave the pool for two whole days except when we took food to him, and brought him in at night. When finally Mimer began to wander around again he did not go far. He would look over the hillside so inquiringly, and when we came out he would ask us clearly to explain: where? Never again, for all the rest of his life did Mimer sit on the mat outside the kitchen door, which he had so determinedly tried to share with him while Kirk lived.

Mimer was having difficulties of his own in the spring of 1965. He had developed a distinct limp in his right leg though there was no swelling visible in any joint. When our family doctor, always sympathetic to our pets, examined his legs, Mimer indignantly protested, swinging his neck from side to side and vibrating almost audibly. Bob decided that Mimer's general health was good and that he would be able to enjoy life more if relieved of the pain that caused his limp. So he prescribed for a limited time, a very mild cortisone, which our family druggist in Denver sent to us. We opened an acount under Mimer's name at the pharmacy. Along with the second prescription, he sent a friendly note:

> I am glad to learn that your duck is improved. My wife had quite a laugh when I told her I was late, waiting for the doctor to phone a prescription for a duck.

> Give my belated congratulations to Mimer on his 10th birthday, and I am happy to learn that he is doing well.
>
> —Homer

Mimer took the pills about as willingly as most children, requiring two people, one to hold him, the other to open his bill and push the pill down far enough that he couldn't spit it out. Each time, as soon as he had swallowed it, taken a drink, and gotten his breath back, he reprovingly lectured us. His energy did, however, pick up immediately. He was able again to scratch himself on his right side, to stretch both legs and both wings again, and to preen as before. But inasmuch as continuation of cortisone would have brought on more undesirable side effects, Bob suggested turning to palliatives: cod liver oil drops and baby aspirin.

Since our mountain hillside did not harbor worms and slugs, the main meat diet Mimer had known at 2049, we thought a can of fishing worms would provide a beneficial, high protein treat for him. We proudly brought some home one day and dumped the can in front of Mimer, as he stood expectantly waiting for the treat he understood to be forthcoming. A great glob of entwined, emaciated, twisting, skinny worms wriggled on the sandstone. Mimer took one look at them, gave a long disapproving quack, and backed off. Even after we disentangled one or two and held them up temptingly, he refused to have anything to do with such undernourished, sickly specimens.

But we did happen on a substitute which became a source of delight to Mimer. A good friend from Denver who came each week to have Sunday dinner with us had brought a steak, as was his generous custom. We cut some raw pieces into Mimer-sized bites, and put them in his "goodie" dish, holding a conversation with him while preparing it, so that he knew a treat of some sort was in store. It turned out to be exactly what he had been looking for! Mimer's enjoyment had been so keen that the next Sunday when Byron came for dinner, we again took off a piece for him. There after Byron's very arrival of a Sunday, at the front door downstairs, brought Mimer to the kitchen door upstairs, clearly announcing that his meat had arrived, and could he have it. His uninhibited frankness was so appealing we forgot to be embarrassed. What was significant was his clearly deductive reasoning that Byron's arrival meant the steak he so enjoyed had also arrived.

Moulting became a longer, more uncomfortable process each year, the new feathers showing striations similar to those in human toe and finger nails as age increases. Mimer rose later in the mornings than before, and merely sat outside the kitchen door to wait for his egg. He was more willing to stay inside quietly, and he dozed more and more frequently during the day, his bill dropping until it touched the floor, abruptly waking him. Like an old man falling asleep over his newspaper, he was momentarily embarrassed at being caught napping, and we pretended not to have seen.

It was during these days in the kitchen, after Kirk had left us, with only Gobi above in her open-door cage, that our mutual depth of understanding grew yet more. Mimer and I were two fellow creatures, co-existing, needing each of us the warmth and reassurance of the "life intelligence" in the other. I knew—and perhaps Mimer did too in his duck comprehension—that we shared a tie that could not be broken, between two manifestations of the great life force.

Because our uncertain Rocky Mountain spring seemed to make its appearance early in 1966, we tempted the winter fates to return by filling Mimer's pool in March. He slid in as soon as possible, and splashed around with much of his old enthusiasm. But he did not stay for hours that day in the delightful freshness and novelty of the first swim of the year. He clambered out with obvious difficulty within minutes, very wet indeed, preened, and then waddled, limping, over toward the house. He did not go back to enjoy the sensation all over again. This unusual behavior continued for several days. We discovered he had developed a sore spot on his increasingly prominent breast bone, almost in due center of his body. With his stiff leg, it was increasingly hard for him to clamber out of the pool, and in the climbing he therefore scraped the bone. Moreover, his feathers seemed strangely unable to shed the moisture—he became water-logged. He finally refused to go near the pool at all, and ignored its very existence from that time on. To realize that one of his chief joys in life was gone was profoundly sad, but he appeared to have lost interest in the whole thing, which was the most sensible attitude he could have assumed, and quite in keeping with his general adaptability.

1966 brought Mimer fame in Boulder. A friend who had known him in Denver, and had been in the P.T.A. audience when he and Jemima performed on the stage ten years before, upon learning that Mimer was still with us, persuaded me to write an article about him for the local paper. The result was that he was featured in full color on the cover of their Easter Sunday rotogravure section.

After the story had been accepted, we tried to prepare him for being dressed up in Jemima's old costume which had been tucked away for so long. The scarf he tolerated with a fair amount of cooperation, though very briefly. The bonnet, however, as it always had been, was taboo. We no sooner got it on his head, and the ribbon into a semblance of a bow, when he would give one quick flip and it skidded down his long neck and rested upside down on his breast. Once in a dozen times he would hold still for a minute or less.

When the photograhper from the *Boulder Camera* came, the scene to Mimer was merely a repetition of the foolishness to which he had been subjected during preceding days, with the added annoyance of this stranger and his alarming paraphernalia. By dint of Laney's sitting with him, the man finally got a picture which did Mimer fair justice. But I was inclined to agree with Mimer that he would have looked just as appealing in his natural attire.

An important event resulted from the publication of Mimer's story: Mimer's meeting with Mabel, a nurse at one of the local hospitals, a kindly, mature woman whose childhood had been blessed with an abundance of pets, among whom had been a duck. Mabel had read Mimer's story in the newspaper and asked if she could meet him. We were delighted at her interest, and one morning soon after, Mabel came up the canyon. Mimer was in his corner of the kitchen. Mabel walked into the room, we conversed very briefly, then she knelt down beside his pen. Since Mimer made no fuss about her familiarity, we shoved aside the cardboard wall so that there was no barrier between them. Even though Mimer liked women better than men, he was sometimes afraid of strange females. Normally he would have retreated toward the wall, there to carry on whatever conversation was required.

But without a moment's hesitation Mimer not only pudded toward Mabel, as though he fondly recognized her of old, he stepped up and into her lap! She sat down to make him more comfortable. Mimer contentedly settled down, as she stroked and petted him and talked at length with him.

If I never see Mabel again, she will remain always in my memory as a unique and loving human being, who called forth from diffident Mimer a response that travelled beyond his experience and acquaintance as well as hers. We were onlookers at a very mysterious meeting, fraught with the significance of a recognition, a previous acceptance, a trust, an intimacy.

Shortly after Mimer's remarkable meeting with Mabel we noticed that he was extremely uncomfortable and restless during his long hours of sitting, and we realized that his tender breastbone suffered against the pressure of hard bare earth or the prickly, fallen pine needles around the house. With belated insight, we saw that what he needed was a soft pad to rest upon. It had to be washable, without destroying its buoyancy. We made three foam rubber pads for outdoors, the kitchen and bed. Mimer was somehow way ahead of us and knew immediately what he was supposed to do when we introduced him to the outside pillow. He climbed right up on it, and his ease and pleasure were so evident in his contented small quacks as he settled himself on his cushion, that we were ashamed not to have seen his need before. He adopted the other two with equal assurance, and used them for the rest of his life.

Mimer's baths were limited forever now to a bathtub. He told us when he wanted a bath by ducking his head in his water pan and bringing up that remarkable load on top of his neck and scattering it over the room. After we had pulled the shower door across the tub, providing him with the privacy he required, he splashed and swam around and did as well as he could in the one-shot operation, minus the rinses which were so much more satisfactory. But we soon heard insistent quacking from the bathroom, as he felt himself becoming water-logged.

With Mimer's increasing difficulty in walking, he spent most of his time on his pillow next to the glass wall of the dining room. Perhaps he saw Jemima reflected as he had in Denver. Whenever we stepped to the window to check on him he quacked recognition and reply.

Not one sign of age which appeared in Mimer ever showed a semblance of senility. In fact, his lameness was the chief indication that he was slowing down. He made less frequent trips to the kitchen door to ask for his grain, but waited to have it brought to him as he sat on his pad at the dining room window. One morning, after he had eaten his egg at the kitchen door he made no effort to waddle the dozen feet or so to his pad, but sat—plunk—on the ground. He obviously found it too painful to walk. As I lifted him to his feet, with my hand under his broad keel, an idea clicked simultaneously. His neck was pointed in the direction of his pad, and with my hand supporting him like a crutch, he paddled his feet vigorously. Together we moved as fast as he had been able to do as a young duck. He clambered onto his pad, then turned, and seemed to tell me either "Thank you," or "Why didn't you think of this a lot sooner?" Or perhaps his quack said both. After that, Mimer called when he wanted to move; I came out, placed my hand as his crutch; he pointed with his neck and head in the direction he wanted to go, and we arrived smoothly.

On March 3, a Sunday morning, I lifted him off his bed-pad, because he was getting so stiff he could not get up alone in the mornings. We had our usual quiet "mushing," with his head buried in my lap, and soft mutterings on his part as well as mine, and I carried him upstairs. The flesh under his keel felt rather jelly-like, a condition I noted but made no point of at the time.

He asked for his egg when I set him down outside the door. And after eating it, he requested help to his pad. He stepped onto it with assurance, and plunked down, completely without shock absorbers these days. Toward the middle of the morning, Laney went out to take his seeds, and found him unresponsive, as though he did not hear her or know she was there.

Whether he was then aware of us or not we will of course never know, but we sat beside him, absolutely helpless, as life fought to hold on and the forces of physical disintegration proceeded relentlessly. It seemed an eternity, but was probably only a matter of an hour, that his labor to breathe ended, and with a shudder his little body stiffened—and we covered him over.

Today his ashes repose next to those of the dog companion he loved so long and so well. But he himself still resides with us, fresh and real, though he is now the free one, and we, for yet awhile, "prisoners of the splendour and travail of the earth."

Mimer had been a part of our thoughts, our everyday care, for so long, that when he left, the emptiness, the hollow, the utter goneness was excruciating. I kept listening automatically for his quack outside, needing us. Toward the end he was fully as much our care as Pecky.

I don't suppose Mimer was an outstandingly brilliant duck. But then there are very few human animals who are brilliant either. That does not make him or them less interesting, or significant. His basic nature was shy and retiring, and it required gentleness, a long time and steady encouragement for us to develop rapport and to secure his hesitant affection. Perhaps if he had not suffered the loss of Jemima he would never have felt the need to respond to our efforts.

To peer only at the physical properties, magnifying a feather, distorts the image of the whole. It was the mysterious life force which molded both Jemima and Mimer, made them the individual entities they were, possessing each their own intelligence "not to be measured by man." Nor can we ever measure the intense joys, the breadth of education which Jemima offered in her two short years, that Mimer gave his family by living with them for thirteen wonderful years.

Chapter Twelve

WIGGA

In contrast to Pecky's quiet and somewhat prosaic life, Wigga's was a meteoric career, and sadly about as short. It seemed as though everything he did, he did hard, and with much fanfare. Pecky's gentle, persuasive ways were alien to his disposition. He demanded and battled, scolded and otherwise tried to intimidate, to get his way. Of course he had never suffered physical injury or permanent handicap as had Pecky, and was in the most robust health all his short life.

Even the premature fall from his parental nest, down by the railroad tracks, and being subsequently stuffed unceremoniously in someone's pocket for hours had not done him any noticeable damage. His survival strength had been enhanced genealogically by centuries of ancestors pitting their wits against the most wilful and malicious mistreatment by humans. (If I were a pigeon I would get as far away as possible from anything that had to do with man.) So no one thinks a pigeon is extra special. Particularly a mongrel pigeon without a pedigree or the status of being either a homing or a carrier pigeon.

Our neighbor was the night dispatcher of the Denver and Rio Grande Railroad, and worked across town at the railroad yard. He was a kind man, who loved and was loved by our Kirk, and even by Mimer in his own diffident fashion. Johnny could not bear to leave in the street, to be run over perhaps, that poor, unprotected, abandoned baby bird, no matter how fat, or ugly, or common he was. And so he stuffed him in the only container he had, which was his pocket, and brought him home to us. The yellow baby fuzz was still clinging to the tips of his first dark coat of feathers, giving him a wispy, frowsy look. His orange eyes were big with fright, and his over-size, greyish feet spread flat over our hands. He was very ungainly, in fact rather ugly, with a bump on his beak (the hard protrusion which had enabled him to break open the shell which had encased him) where it joined his baldish head. This made him look cross instead of fearful, which he was.

Even for so young a pigeon he seemed to be very large, weighing almost twelve ounces, whereas Pecky had never topped four at the height of her maturity. His baby cheep was as loud as he was large.

Wigga

Loud and insistent. He cheeped about everything, whenever we came near, whenever we offered him water or food, whenever we left him alone, whenever he was hungry, which seemed like most of the time. He also wiggled his wings whenever he cheeped.

He was so young that his beak had not yet attained its proper hardness, nor had he learned yet to use it to pick up his food by himself. This meant that we had to pry his bill open with one hand and push the moistened grain into the sides of his bill with the other. As it was indeed a strange method of being fed, he squirmed and struggled to be free of our clutch, cheeping loudly all the while. But this particular baby bird had a long heritage of unusually harsh living conditions, from ancestors who had persisted in the most unhealthy, dirtiest, and leanest of environments, the railroad yards of any city.

He was truly a most unattractive fledgling, whose potential only a parent could believe in. He did not know of course and could not guess, how far he had to go to be acceptably attractive. All he exhibited then was an enormous appetite and a determination to have it satisfied. Laney became his surrogate mother, and in a very few lessons Wigga was mastering the technique of spotting a seed, coordinating sight and muscle by opening his bill over it, pecking it, drawing it into his mouth—and he was off to independence at mealtime. He found Mimer's seed pan expeditious and helped himself as readily as if Mimer had invited him to do so.

He showed no desire to be independent otherwise of the family with whom he had come to live. He early found the top of the kitchen door opening to the utility room an excellent perch from which to observe our activities, and he would sit there like a wise commentator, bobbing his head, muttering his pigeon talk, and, every time anyone looked at him, wiggling both wings vociferously. Pecky's tentative and infrequent flutter of one wing at a time seemed a whisper of communication in comparison. In fact, the wing-wiggling became so integral a part of Wigga's demonstrations, so much his hallmark, that his name dropped on him because of it. Wing-wiggling becamc his signature, his identifying idiosyncrasy all his life. While "vociferous" refers by derivation to the use of the voice, its antonyms listed in Webster so aptly describe the opposite of Wigga that they help to delineate him: "silent, quiet, calm, subdued, tranquil, sedate," not one of which he ever was.

He insulted everyone gaily and without reservations, a most uninhibited character. He sat on Kirk's head or walked along his side if he were lying down, dipped into Mimer's food at will, followed the big white duck around, blissfuly unaware that Mimer either ignored or scorned him. He scolded and chased various members of the family off of territory or out of rooms he had appropriated, and insisted months later on making Laney play a certain role he had decided she should. He did not hesitate to bluster up to a guest and try to chase him or her out of a chair or a room. Pecky simply retreated out of sight if she disliked or distrusted a stranger. Mimer scolded from a safe distance. Kirk and Kerry were very proper, cool but courteous. But Wigga was undeterred by niceties or by any sense of propriety. Zooming over someone's head was merely the way one got to where one wanted to go, and no area was sacred. He learned the lay of the house very quickly, but, surprisingly, seemed to accept one limitation. He was forbidden the living room. He would somehow call a halt to his brazen assumptions at the doorway, and merely stand there. I almost gave in.

The first few months of his life were spent in Denver, where Wigga quickly assumed his territory, which seemed just about to coincide with our lot lines, and he flew gracefully around the outside of the house, coming to perch on any of the window sills, but especially on the one outside Laney's bedroom, where he could see her working at her drawing board. Our window ledges were not the easiest things on which to keep one's balance, for they were of bricks laid on a slant, and Wigga was constantly slipping and having to reinforce his footing. He would land with a thump, and notwithstanding their slant, would begin his round and round dance, fluffing his neck feathers at whatever stage of moult, quill or paintbrush, and coo loudly. In a few weeks he ventured to the roofs of the houses on either side of us, during which forays he met pairs of his own kind to whom he reacted not at all, other than to ignore them, and to move away if they approached him.

We had some anxiety that September as Wigga disappeared for two days. We wondered if he had returned to the nest where he had been born, which we had been told he would do. But on the third day he reappeared, chipper as ever, his adventure, wherever or whatever it was, having done him no harm.

As Wigga gained maturity Laney thought it an excellent idea to build him a house. But it was soon obvious that Laney's idea of appropriate lodging did not coincide with Wigga's notions of a bed. He confused and misled us at first by his enthusiasm for the building of the little A-frame she designed. Wigga pranced over it and over her, gronking his pleasure at being a part of the action, and strode into the front end, strutted over the floor, and flew out the back.

He demurred a little about going in when she put on the door which caused the front opening to be smaller. But he balked absolutely when she nailed on the back, thus blocking his rear exit.

She tried shoving him in, but he became a feathered fury and flew out past her and up to the top of the clothesline pole, there to scold roundly. No amount of coaxing, putting food inside, or any other inducement could change his mind.

Wigga participated in all outdoor activities of the family, never doubting for a moment that he was welcome. Through the late spring and summer he was present with gusto at all picnics in the backyard, and accompanied whoever mowed the lawn, alternately flying, walking in front of the push mower, or riding a shoulder. When we put our house up for sale in October 1963, preparatory to moving to the mountains, he saw nothing unusual or unacceptable about asserting himself to every prospective buyer. He could never have known how much of a deterrent he was to any sale at all. He invariably accompanied the real estate agent when she took clients around the yard, his insensitivity beautifully unaware of the fact that she could barely put up with him. If her commission had not lain in the balance she definitely would not have! Of course, we had to admit it was disconcerting to have a pigeon land on one's newly coiffed head! In fact, we had to assure the woman who purchased 2049 that Wigga did *not* go with the house, as she unfortunately had no love for animals.

On the actual day of moving, Wigga was simply frantic as he saw his familiar landmarks disappearing. He flew in and out of the open doors, followed the movers, got in their way by flying over their heads and generally made himself a nuisance. It was pathetic to see his insecurity.

As it turned out I think Wigga was attached to us by that time and not to the house. He adapted with alacrity to the tiny cabin high on a hill where I finally found refuge with our family assortment. Wigga took note of the situation and promptly set about determining territories and boundaries, his own and everyone else's. The cabin was so small that the outlines of what he wanted were rather quickly apparent.

We had been told that Wigga would never remain with us in a new location but would return to his origins. However, at 2049 he was now unwelcome; and the railroad tracks were not acceptable to us even if he could remember where he had been born. He was by that time a much loved member of the family, and I had no intention of losing him. To turn loose an animal to fend for itself, who has become used to and dependent on our care, I will not do.

So during his periods outside we put Wigga into the wire enclosure erected primarily for Mimer's safety. Neither Mimer nor Wigga thanked us for our provisions and felt instead that we had done each of them wrong. As for Wigga, the wire enclosure merely put him within reach of Mimer's resentment and his only refuge became the window sill to which he had to retreat where Mimer could not reach him. Otherwise Wigga spent his days more happily inside the cabin.

He chose a small book shelf high on the wall of the miniscule living room and there paraded over my Shakespeare and a small dictionary. It was the closest he could come to a perch comparable to the kitchen door in Denver. When he tired of that he either zoomed or strode purposefully through the kitchen and into the tiny bedroom where for some reason obscure to us, he decided to take up quarters, which meant that it was off limits for me. He would come rushing out from under the bed, gronking, and literally chase me out, pursuing me as far as the door. There he would stop and stand guard. It was no game as far as he was concerned, but a vital confrontation. It was perhaps Wigga's lack of a sense of humor that made him so funny. Life was a very serious business to him even though he was not bowed down with the weight of a constant fight to obtain food and shelter. His domestication did not follow Pecky's path at all, nor admit of leisure in which to play, as we watched her do, and as our other birds have done.

At night we housed Wigga in the cage in which he had travelled to the cabin and sat it on the board next to the tiny sink. As he grew to his full size we almost had to stuff him into it. He resented the confinement, and even more its inadequate size, and would wake everyone in the middle of the night by deciding to shake, which rattled his cage and woke Mimer to a cross, complaining quack. I discovered, in our close proximity there, that if I turned over in bed, separated by six feet and a thin wall, Wigga took it for a signal to gronk his displeasure and to ask for release from his box. This would be followed, when he failed to rouse anyone, by another violent shake. Sometimes Kirk would come from his rug next to my bed, lay his head on the bed, and inquire what was to be done about the commotion in the other room. If I answered in anything louder than a whisper, Wigga really went into action, provoking Mimer to louder objections. Then Gobi would begin to flap in frightened confusion under her cover.

The miniature size of our cabin meant that Wigga exhausted its possibilities very soon. And it sometimes seemed that out of boredom, he hunted up mischief to get into! One cold and blustery morning I was particularly busy, and having both Mimer and Wigga loose in the house was two birds too many. Suddenly there was a horrible clatter in the tiny living room. Immediately Mimer and Gobi stopped their chatter and were as silent as the walls, waiting to see what would happen. I came to the door, and there sprawled all over the floor were the contents of my sewing basket, and Wigga chasing one of the rolling spools of thread. He had landed on the side, apparently, with his usual thump. As the basket had spindly legs, and was rather delicately balanced, he had tipped it over.

I began to pick up the scattered mess and scolded him roundly. I just happened to look back and saw Mimer peering round the corner, just like another child waiting and almost hoping for punishment to descend on a resented brother. Wigga was gloriously unconscious of having caused any difficulty, and even had the audacity to peck at me as I picked up after him.

The cabin did have electricity, so I could use my Electrolux. Wigga showed no fear of any household appliance no matter how noisy it was, contrary to Pecky, Mimer, Kerry, and Kirk who were all distressed and alarmed by the vacuum cleaner, its noise, and the long hose. But Wigga took an immediate fancy to the reflection of himself in the shiny metal disk at the exhaust end of the motor. He hadn't liked his own kind in the flesh in Denver, but this was an undemanding sort of being. The swishing of the hose and the roar as I cleaned, bothered him not at all. He flirted outrageously before the lovely pigeon he saw there, puffing out his iridescent neck feathers, gronking, and going round and round in circles. As I pulled the cleaner along, he followed.

It was therefore an interesting encounter months later, when he refused and repelled any and all attempts of a real pigeon to make friends. He had by then installed himself at Dunyana and Dale had nailed two perches for his use just outside the high fixed panes of glass above the glass doors in the kitchen and in the den. These shelves were protected from weather by the deeply projecting eaves; they also permitted Wigga to look inside and to gauge which direction we were headed if we left either room. One snowy evening that following autumn, just before it was time to bring him in, we heard a commotion on the perch outside the kitchen. Going out, we looked up to see Wigga scolding and occasionally flipping his wing at another pigeon who had materialized out of nothing, and who also appeared to be feminine, judging from her actions. She was at one end of his perch, obviously trying to make overtures to him. She edged along the shelf, and every time she did Wigga would slide away, until the perch gave out and he flew to the den perch to get away. She followed and the procedure was repeated. He appeared positively scared of or repelled by her, and rejected her more roundly than Mimer had rebuffed Ducky a few years before. Wigga gratefully flew to our shoulder and came in the house. He apparently preferred a narcissistic affair with his own image in the cleaner than one with a flesh and blood reality. Or had he become such a people pigeon he could no longer relate to his own kind? The poor visiting pigeon, coming from we never knew where, was gone by morning.

Having had to cohabit with Mimer in such close quarters during our cabin-stay Wigga no doubt felt there was an added intimacy now which gave him privilege to continue in the same vein. In the first early spring at Dunyana, when Mimer was wading around in his makeshift pool, frustrated because it was not deep enough for him to get his feet off the floor, Wigga frequently hopped on the low and wobbly wall and then jumped in near the edge, quickly trying to squat and dust himself with water. Mimer was cross enough without having an unwelcome intruder, intimate or not, so he would steam over, his neck straight out and his bill open, to nip Wigga. The importunate pigeon fled, but would flutter back undaunted and in due time try again. After a few repulses he lost interest and usually flew back to his shelf, there to do a go-round or two, and gronk disapprovingly.

With the building and filling of Mimer's cement pool, Wigga assumed, being Wigga, that it must be for his pleasure as well as Mimer's, even though he still did not know how to make efficient use of a pool of water. His powers of observation were keen, and from a distance he watched Mimer, in a transport of joy, dive and splash and bathe, throw the water all over himself, and make racing runs from one end to the other, quacking and mumbling to himself. So Wigga decided to march importantly, and rashly, over to the pool. He stood at the side where the slope was least, and eventually stepped down and into the water. Whether he ever would have learned to take the thorough bath which the juncoes delight in, or the robins throw about, or Gobi enjoys as she soaks and gets soaked, is a question. Perhaps he thought he might even swim like Mimer!

If Mimer had run him out of his pool, which he did rather often, I would provide Wigga with a large pan of his own. But I am not sure that he ever really understood what a bath was all about. He would almost tip over on his head in the effort to get his front wet, but forgot about the rear or his back. Yet he was a clean bird about his person, and never to our knowledge had a flea on him, even when Johnny had brought him home to us.

Wigga's numerous thermal layers of feathers kept him warm into quite wintry weather, but he, like all creatures who have known greater comfort, chose not to stand the rigors of the cold, and asked to come in if it snowed. The innermost feathers were so fuzzy and contained such a charge of electricity or magnetization that if we held a detached one it curved around our fingers and had to be pried off.

Wigga never objected when we held him upside down and explored the depths of his feathers. Such an act was abhorred by Pecky, as an invasion of her privacy and an indignity, and Mimer felt equally as strong about it; Gobi retaliates with a good bite. But Wigga was so glad of attention that he would put up with anything, even if he had to get it upside down. We could pick him up at any time and place, affectionately squeeze him, ruffle his feathers or bury our faces in them, and he would gronk happily all the while. He gladly straightened things out afterward with a tremendous shake which all but knocked him off his feet, followed by a prolonged, cooing preen. Wigga had achieved what he most wanted—attention and love. Our other birds have impatiently felt it necessary to voice their displeasure and to show it by getting rid of the "human" as thoroughly as possible, either with an immediate bath or by a dry preen.

As spring and summer greened our hillside, Wigga's observation posts under the eaves became more a means of knowing where his family were than to scan the scene over the hill. The shape of Dunyana is a wide "V," and his post over the kitchen door was at the base of the "V." As he peered through the glass he could tell if we went to the west wing or to the east. Without fail he would present himself outside the glass door of the proper room, going round and round, gronking, demanding to be let in. If we slid the door open he would stride in with all the assurance in the world, to take over activities as he saw fit. As soon as the weather became warm enough we began the custom we have followed every year since, of sitting, after breakfast, on the sheltered east side of the balcony which surrounds the living room.

Wigga's scouting duty informed him where we were, and we would no sooner get seated than he came sailing around the corner of the house, landing on the railing with a thump that jarred the whole balcony. He was so pleased with himself he strutted and puffed his neck feathers and went round and round, cooing vociferously all the while. Woe to the coffee cup that sat on the railing, for Wigga was oblivious to impediment and cared not if he knocked it to the ground beneath. Kirk would look at him with bored disdain, and affected not even to see him when Wigga flew down to the floor of the balcony and paraded in front of him, or committed the final indignity of making love to Kirk's nose. The big dog, with characteristic tolerance, merely tried to ignore the whole thing. It never occurred to Wigga that the silence of the hillside which we craved, and for which we went out early in the morning, was shattered by his noisy effrontery. Nor could he be persuaded to keep still. Because if we grabbed him and loved him, he gronked even deeper and more incessantly. So different from gentle Pecky, whose quiet, earthbound explorations of backyard or hillside were carried on as unobtrusively as her blending grey coloring, and so silently we had to keep an eye on her in order not to lose her. Wigga could not for the life of him, ever have done anything gently or quietly. He was simply not built that way.

Laney was that summer finishing up some courses at the University, and on a few hot weekends chose to do her typing outdoors on the cooler, north-facing downstairs terrace. Wigga from one of his observation perches spotted Laney's departure, immediately took off and came gracefully wheeling around the house to find her. He swooped to the ground announcing his triumphant arrival with parading and loud cooing. Then he flew to her shoulder to observe this fascinating new amusement. As her fingers made the old manual machine click and rattle along, he had to get in on the act and hopped to the typewriter itself, landing precariously on the keys. He hopped to the roller, back again to the keys, walked along them, which required no small feat of balancing, and tried as she tapped a key to catch it en route to the paper. As usual, he was not to be rebuffed by a suggestion, that being too subtle for his direct nature. Tossing him into the air only fortified his determination. The only solution was to retreat indoors, and leave Wigga in impotent frustration, doing his dance outside the glass door.

Wigga always came indoors at dark along with Mimer. He would begin to scold to be let in as the light failed. When we slid the glass door open he would already be on the ground, and would march in imperiously, so obviously important that Laney began calling him "Mighty Mouse." But whereas by 8:30 or so Mimer began to be tired and cross and wanted to be put to bed in his corner of the garage downstairs, Wigga was game to stay up to the last man. As with an obstreperous child we welcomed the hour when we could demand his retiring. Once he was particularly obnoxious and I put him to bed at 4:30 in the afternoon, but I paid a price. He was making a clatter by 5:00 a.m. the next morning.

From among the other animals Wigga chose most determinedly to try and identify with Mimer, whether he realized or not that the big white duck was another bird. Gobi and Wigga ignored each other's existence except for the infrequent occasions when Wigga flew to the top of Gobi's cage. If he stood there long enough Gobi would take aim and give his foot a vicious snap, a hurt not to be lightly dismissed. Wigga understandably jumped straight in the air, gronked protestingly and flew off. This ended the interaction until the next time Wigga accidentally landed on her cage. Gentle Kirk was part of the furniture of his surroundings, whom Wigga showed no fear of whatsoever, even when Kirk occasionally sniffed him over thoroughly. He merely used Kirk as he used everything around him, for his own purposes. Mimer was especially long-suffering on this score, having to endure sharing his grain dish, his lettuce, and his sleeping quarters, and even having Wigga tag along after him.

When Wigga joined the family he entered without question into the morning ritual of the hardboiled egg, and with characteristic decisiveness. It was the discipline about when and where he could come into the act that forced him out of character. He had to learn the hard way and by being forcefully repulsed, that Mimer got his share first. But the waiting almost undid poor Wigga. I used to think sometimes, Wigga could have survived in competition with the unmannerly and aggressive ruffian pigeons in the Piazza de San Marco in Venice. But that is probably unfair to Wigga.

Mimer seemed to us to charge into his egg and almost swallow it whole. He did in fact get too big a bite sometimes, and would have to swallow hard and wait for the bulge to make its way down his long neck before he could contemplate another, meantime looking rather dazed. But his broad bill was soon at it again, scattering crumbs on the ground under our hand. This was the portion Kirk stood by waiting for, drooling if Mimer took too long, wagging his bushy tail in happy anticipation. These meagre crumbs could by no stretch of the imagination have satisfied any physical hunger; it was mostly a psychological necessity. The same had to be true of Wigga, for when he got his share it amounted to even less than Kirk's. He had been firmly ousted by Kirk from the ground crumbs, and had taken up as his waiting post a pine stump that sat by the kitchen door. Thereon he danced his round-and-round, scolding everyone. When Mimer had had all he wanted of his egg, there was always a residue of small crumbs left on our hand, which we then offered Wigga, who wiggled his wings mightily as we turned to him. His excitement made him rougher than usual, and his sharp beak was not a little uncomfortable, so we rubbed off the larger crumbs and left him to peck them off the stump, which he did with rumbling accompaniment.

Since Wigga minded not at all being ruffled up and handled by us, it probably never occurred to him that anyone else could object to similar treatment. Mimer was mostly a non-contact person; Pecky requested at least a certain amount of restraint; and Gobi is even more of a hands-off character; but Wigga was distinctly a contact being. He did not mind climbing over any and everyone. Nevertheless, in spite of his brashness, we found him almost always irresistible, and forgave his inability to refine his interpersonal relationships. He really did become a beautiful physical specimen of a pigeon, whose feathers were thick, healthy, iridescent and attractive in color. His tail was a dusky navy blue, his long primary wing feathers white. His neck was also dark but filled with the shimmering jewel shades of the male pigeon. The remainder of his coloring was a light, blue-grey, with a wide band of darker hue just beyond his middle section. It was no wonder he liked to admire himself at the back of the cleaner or anywhere else that he found his reflection.

The longest sustained performance Wigga ever put on was throughout his second summer in 1964. As with most of his activities it was both funny and pathetic. But also like most of his antics he pursued his purpose so doggedly and without surcease, so single mindedly, that we who had to live with him sometimes lost the humor of it and were more annoyed or provoked than as sympathetic or understanding as we should have been. And I might add, with longing hindsight, as we now wish we had been.

Wigga was so very earnest, and wanted what he wanted so intensely. He knew only one way to get it and that was the straight, ramrod approach; no subtleties softened his nature. Hence, when the idea hit him about the middle of June, he went to work immediately. We were sitting on the balcony where he had joined us with his usual flourish, wheeling gracefully around the corner of the house, landing on the railing with the accustomed resounding thump. He immediately paraded, round and round, on the narrow railing, his rumbling coo as penetrating as usual. We had not yet been able to afford deck furniture, and were sitting on some old pillows, on the floor of the balcony. Laney's was a nondescript, square grey one, with a large round yellow button making a center depression.

She got up to get something inside, and as soon as she did so Wigga flew down to the pillow and ensconced himself delightedly, fluffing out fully over the yellow button. He absolutely would not leave when she returned, so she lifted him off as he protested loudly, and tossed him into the air. Nothing daunted, Wigga flew back and tried to bluff her off the pillow, bustling up and pecking at her, flipping his wing sharply against her leg. He seemed to be even more serious about his idea than usual and refused to be sidetracked.

Finally he flew to the ground and back up to the balcony, bearing in his beak a stick which he presented to Laney. Suddenly light dawned on us. Laney moved to the side of the pillow and Wigga made another trip down to the hillside. When he came back he had another small stick and a pine needle. Seeing the pillow vacated, he promptly desposited his mouthful in the middle, neatly over the button. Then he hopped onto the pillow with 110% concentration and settled himself down, fluffed around the sticks and the button, every bit as efficiently as any hen. He leaned over to pull the button into better position, as a hen will use her bill to poke her egg further underneath her or to turn it over. And there he sat, looking so very contented and satisfied. He had achieved his objective.

When we had finished our coffee and prepared to come inside, Laney picked up the pillow, dumping the sticks. Wigga was distraught. He insisted on coming in, following Laney, and when she threw the pillow on her bed, Wigga immediately flew to it, protectively, and settled himself once more on the button just as mother hens settle themselves over their eggs after taking a respite in the yard to eat. Here he remained, happy, quiet, wiggling his wings and cooing when we came in to see him, but otherwise refusing to leave his chosen occupation. All morning he sat on that pillow and at midday he rose, stretched, walked off and came to hunt Laney, where he made himself a nuisance, trying to get her attention. He had something very imperative on his mind that he wanted.

Finally it occurred to us to try an experiment. What if Laney were to go and sit on that pillow? She did, to Wigga's joy and satisfaction. He thereupon threw up the barriers, or in other words established his territory. It did not include my presence. When I came into her room to observe, it made no difference whether he was sitting on the pillow or had succeeded in getting Laney to. He flew at me, flipping his wings, bustling after me when I retreated, scolding the whole time. If I pretended to be afraid he was visibly more excited and pleased with himself. After all, what satisfaction is there in vanquishing a spiritless intruder? Once having pushed me through the door, however (just as at the little cabin), he had accomplished all that was necessary and he stopped. The edge of his territory was well defined and that was the limit of his concern.

For about two weeks following this initial establishment of his nesting arrangements, Wigga sat on that pillow faithfully, every morning, and continued to bring sticks and pine needles to lay over the button. And every afternoon he considered that his half day had ended, and it was Laney's shift. She must have been a severe disappointment to him for she did not often carry out her side of the bargain as he conceived it.

That big yellow button had apparently set all of Wigga's strong reproductive instincts aflame. Laney was accorded the distinction of having produced it and, ipso facto, bore half the responsibility for the setting. Wigga's shift, he had demonstrated, was the morning, in addition to supplying, since she didn't evidently understand about nesting materials. The least she could do was her share of the setting in the afternoon. And until he succeeded in getting this idea over he was truly frantic. It was pathetic; but it was also slightly irritating, for us.

During the "incubation" period he faithfully carried out the orders issued by his instinctive wisdom; he sat on the "nest" for his prescribed hours, and saw to it, or tried valiantly to, that the rest of the time was properly cared for even though his family was so lax and uncooperative. He became so obnoxious sometimes that after Laney had stood all the tormenting she could, she would try to catch him to throw him outdoors. That canny pigeon knew he had passed her limit of endurance, and like a naughty child, flew to his pillow where he sat quietly, "I will be good!" So earnest, so determined, so honest, so very loving—Wigga. As Laney commented, "Wigga wore his heart on his wing."

One day in February, over a year after we had moved into Dunyana, Laney came home for lunch, parking her car on the road above the house. Wigga as always was delighted to see his openly acknowledged favorite, flying up to meet and escort her to the kitchen door. When she left to return to work, Wigga flew ahead of her, landing on the hood of the car, dancing his round-and-round, and cooing with his iridescent breast feathers fluffed out masterfully. Laney shooed him off, started her car, and drove away. I had returned to my work and did not watch her departure.

An hour or so later I went out to check on the animals and to take Mimer's seed dish to him. Both Kirk and Mimer met me with their individual enthusiasms but there was no rush of wings and frantic cooing. I called, but no Wigga appeared. I went out again in a few minutes and called again. No eager wheeling in, no delighted response to the attention he craved. Just that awful blank. All afternoon I spent calling and calling, walking over the hill, to see if he had been hurt and had fallen somewhere. There was not even a bunch of feathers to indicate a struggle of any kind. We sent out SOS's to friends to be on the lookout for the grey and blue pigeon. We scoured the west end of Boulder where our canyon empties into the town, calling for him, stopping to speak to any pigeon we saw, much to their surprise. We were fearful he might suffer a most agonizing death from the poisons being set out by "city fathers" who preferred the dingy, staining soot from the utility company's smoke stacks, and the noxious fumes of diesel buses to the natural guano of the birds.

We even called Johnny who had brought Wigga to us and was still living next door to 2049, but he and Olga never saw a pigeon flying around to the different window sills, or sitting on the roof. Some ornithologists claimed that Wigga would return to 2049 where he had grown up, but it seemed as though he would have left us long before had this instinct impelled him. Others thought his strong reproductive urge would lure him to Boulder, there to find a mate and raise his families. But his reception of the female who had tried to make his acquaintance on the shelf and from whom he had fled made this seem improbable. The button had appeared to satisfy his needs.

The final theory advanced by a friend who had lived in the hills for a long time, was that Wigga had unwittingly been an easy target for one of the many hawks in the area, a sharp-shinned or a chicken hawk perhaps. This was a most disturbing thought, but as the days and weeks went by and no trace was ever found of the energetic, demanding, affectionate little figure "with his heart on his wing," we inclined to the last explanation of his sudden and complete disappearance. In some ways, it would probably have been the least painful of all ends, for if in his soaring over the canyon, the hawk had attacked in mid-air, Wigga would never have known what hit him and the end would have been mercifully instantaneous. But it was so hard never to know.

His going left a wide and deep hole in our little circle. It was a different kind of ache from the one we knew when Pecky left us. Wigga left a gaping chasm, compared to the bleak, unending nothingness that clouded over us when our quiet, gentle, semi-invalid departed. Wigga was the roisterous, imperious essence of hearty, healthy, physical life. Add to that, his equally wholehearted, noisy affection, and it made this silence of his going thunderous and abysmal, and very hard to bear.

I have never been able to look at a pigeon with indifference since Wigga shared his life with us.

Chapter Thirteen

BEE BOP

Our friends, depending on their own philosophical viewpoint, refer to us either apologetically or in amusement as a "birdy family." This is not an inaccurate assessment, since the later animals in my life do turn out to be preeminently a variety of feathered treasures.

I am sure there is something that communicates this "birdy" affinity also to the other species of bird population. Mimer and Gobi were members of our household when the broadtail hummingbird and his injured mate came to our house.

By the time she fluttered to the ground outside our kitchen window, she had clawed the feathers from a portion of her neck, and was weakened from lack of food. Though not the smallest of hummingbirds in this area (the Calliope is a whole inch smaller), she was by far the tiniest bird to which we had ever tried to minister. Her three-and-three-fourths inches included both tail and bill, which was most of her, and she would have cost a half-ounce air-mail stamp to send abroad. Her pedigree was *Selasphorus platycerous*, of the family Trochilidae. But no one could love anything by that name. One becomes attached to a nickname, such as the one we happened on for her, which gathers to itself the characteristics and the loveableness of a creature, the very pronouncing of which will forevermore nostalgically recall that personality.

If our amateur diagnosis was correct, Bee Bop had suffered what would amount to a brain concussion which had blocked or in some way upset the nerve responses that controlled her equilibrium. The painful result, when she attempted flight, was buzzing around in circles, wings fruitlessly beating the air, her feet drawn up, clawing into her neck. Since hummingbirds, even less than the violet green swallows who nest at our house, do not walk or hop, and any movement at all seems tied into flight, she was truly helpless.

When it became evident that her injury was physical, not a case of poisoning from pesticides (which produce similar symptoms), it seemed just possible that she might live if we could find the key to helping her out of her difficulty, and if we were able to give her sufficient nourishment in the process of restoring her to normalcy. We found to begin with, that she was true to her animal heritage of realism, which in our experience appears to dictate that the greater the helplessness from whatever cause, the more complete is their cooperation with the humans on whose mercy they find themselves dependent. It is when they are ready to return to their own kind and environment that they suddenly seem to realize, "Oh my, where have I been! I've got to get out of here, quick." And away they go.

Bee Bop's hospitalization began naturally with her "room." No matter that it was a shoe box, our perennial resource in time of housing needs. Pecky's first shoe box became later her bedroom, as did the one we used for Wigga. I spread an old, soft towel on the bottom of Bee Bop's and her room was prepared.

Convalescent diet was something more challenging. Sugar and water syrup is standard offering for our normally healthy hummingbird customers through the summer season, who use it only for supplemental boosts. As soon as the admission formalities were dispensed with, we offered Bee Bop the syrup with which she was familiar. That it now came from a medicine dropper seemed not to bother her at all, and she drank immediately, as though she had always been hand fed. But in order to provide a more adequate diet for our patient, we held a consultation with some biologists, which resulted in the usual differences of opinion. One was not even interested in her chances of recovery, and wanted only to perform the autopsy. The second suggested the necessary addition of readily available protein to the syrup, in the form of Metracal. When the new concoction appeared in her noon tray, Bee Bop tried it, shook her head, flipped it in distaste and refused to eat. But when the nurse arrived with the next feeding, she gamely thrust her needle-thin bill in, and drank with the dogged determination of a good patient, whose hunger was also a stimulant to cooperation.

As with any bedridden person confined to the walls of her room, our comings and goings made the chief break in her monotony. Bee Bop very soon responded whenever we came near by whirring her wings, with occasionally a tiny cheep for emphasis. No more appealing patient has ever occupied our shoe boxes—she was so very tiny in that expanse of white towel on the bottom of the box, and she looked up at us so earnestly, and with such confidence that we simply had to come up with the answer to her problem. Diet was only a preliminary, a base to maintain health until she could master flight again. And we felt no such assurance as she assumed in us.

Under detailed observation, it seemed obvious that she was trying, in her futile clawing, to connect with something more tangible than the thin air. Her extremely small feet moved so fast and would get so intricately caught up in her remaining neck feathers, that we could hardly find them, and having found them, our hands were so big we could offer no surface small enough that she could grip. Pecky, as well as Wigga had ridden everywhere they wanted to accompany us, comfortably perched on our fingers, but Bee Bop was so many times smaller than they, we began to think of the other birds as gigantic.

A pine twig, the diameter of a match stick, finally answered our need. By shoving it close to her tummy, where her clawing feet would have to hit at some point, we managed to make connection, like the missing piece in a puzzle. Her tiny feet grabbed and hung on, her whirring wings stilled, and she pulled herself up, to sit quite normally on the twig. People who have never sympathetically studied birds at close range, scoff at the idea that they can have expression on their faces. But those who have known birds intimately will know exactly what we mean when we say that Bee Bop's eyes shone with satisfaction, her feathers changed angle on her head, and her whole body visibly vibrated with delight.

Having discovered an initial key, we thereupon moved into physiotherapy, added an additional twig to our equipment, and Bee Bop had regular practice sessions during which she relearned the use and control of her feet by climbing upward from one twig to another, always with an assist from her wings. Each successful climb helped to entrench the newly established circuit, and brought nearer the coordination necessary for actual flight. A setback or two occurred when Bee Bop initiated her own attempts in her box. Once we found her on her back, her neck pulled cruelly out of place by her feet being caught in the remaining feathers. She was buzzing around in circles like a fly on its back, and so we had to insist that she be in semi-darkness with the lid partially over her shoe box until we could supervise her practice.

Bee Bop was not abandoned after she entered our hospital. During the first two days she was with us, the male who had brought her spent long and faithful hours in his self-appointed waiting room on a branch of the pine outside our kitchen window.

To encourage both Bee Bop and her mate, we arranged an actual visiting hour, after she had mastered sitting on her twig. Waiting until he had arrived, we carried Bee Bop out and placed her on a low small branch of the pine. She held on but did not try to fly, and immediately started a conversation, cheeping loudly almost like a call of distress. The male zoomed down, whirring straight to her. Still on the wing, of course, he looked her over, and they became absorbed in each other's presence, forgetting all about us, as we stood back to watch. After some small clucking noises, the male flew to the syrup feeder a few feet away, took a drink, and returned to Bee Bop, hovering back and forth over her head, barely brushing the tip of her uplifted beak with his. She followed each of his motions, her whole body revealing her anguished desire to be able to follow his flight, though she did not yet dare to try. Their talking brought some other visitors, whom her male quickly disposed of, making terrifying lunges at them accompanied by ferocious bill-clacking. Suddenly the male flew thirty feet almost straight up and came down in a powerful sweep within inches of her. In her desire to be with him, to accept his invitation to leave, she rose from the branch—and fell straightway to the ground with a thud.

We picked up our poor, disappointed, hurt little patient, and took her back to her hospital bed. How many falls and disasters that tiny, battered body had sustained. She looked exhausted. The flared tail, with its pattern of bronze and white and deep grey, many of whose feathers were sadly broken and separated, seemed even more tattered and shabby. The lustre of her miniscule feathers, those yellowish-green layers of iridescent scales on her back and head seemed to dim. And her eyes, which had been so bright and eager when with her mate, were glazed and partially closed. The wings, which in the normal flight of hummingbirds appears transparent in motion, but are a greyish brown in repose, drooped at her sides, especially the right wing. Defeat did seem very near.

But given a little time, her marvelous resilience was also near and ready. She needed only more rest, food, more practice and above all, constant encouragement.

Bee Bop's original hospital room was not large enough to

accommodate her physiotherapy. In order to encourage her self-help rehabilitation, Laney garnered a larger box and improvised a gymnasium of sorts, complete with a complex of forked twigs and perches. Bee Bop's second big breakthrough followed at the end of the first week.

One day we came with her mid-morning feeding, and listened for the accustomed whir of her wings anticipating our arrival. But this time, the whirring increased both in volume and in speed, and when she saw us, Bee Bop *lifted herself* off the bottom of the box, and without any spinning round, or clawing of the air, howsomever timidly, she flew to one of the perches and lighted like a feather. Never mind her hanging on like a leech. She had done it! Bee Bop settled in hummingbird fashion, with a short, quick flutter of her wings, and a satisfied flip of both, close to her sides. She radiated the joy of her achievement in eyes which were never more bright, and in the iridescent green feathers on top of her head which she raised in her excitement. In our mutual jubilation we almost forgot the dropper in our hand, until she reached for it. She drank with the most enthusiastic appetite she had yet shown, and when she finished, she wiped her long bill back and forth on the twig, just as expertly balanced as any healthy hummingbird we have seen.

Feeling thus elated, we called another consultation, this time with the famous ornithologist, Dr. Robert Niedrach, of the Denver Museum of Natural History. We wanted his opinion on her prognosis now that she had made her second big improvement. He said, "Keep up the good work. And don't get discouraged if there is retrogression once in a while. Give the poor little thing every chance. And if she finally is able to fly away, watch to see if you can recognize her around. If she survives the first few days away, then it is quite likely that she will make it."

Fortified by this expert reassurance, we doubled our hours on duty. She had refused from the first to have us 'round the clock, and had shown her scorn of a hospital routine that attempted to feed her on a 24-hour schedule. When we mistakenly tried to feed her late that first night, she merely exhibited glazed indifference, and refused to come out of the sort of semi-comatose state into which she quite normally sank at close of day. She had also refused adamantly to try the addition in her diet, of a suggested item of aphids. Either we served the wrong brand name which we had located and gathered with no little trouble, or we misread the directions for serving. Whatever the reason, she seemed as much afraid of them as Pecky once had been with a worm which we presented to her under the erroneous notion that she needed her protein in this form; and as affronted as Mimer had been by the wriggling skinnies we had, with such good intentions, once offered him.

Bee Bop had now advanced from the twigs and perches in her box to the whole house as her area of recuperation and flight practice. She progressed in swift strides from stepping onto her twig when we held it, to flying from it to another perch somewhere, anywhere in the house. She seemed to enjoy flying in the kitchen, and would zoom over Gobi's cage on her way to the window just behind. Gobi found this over-sized bee of tremendous interest, and we were grateful that Bee Bop never actually landed on the cage, because one of Gobi's bites would have done a great deal more harm to her than they did to Wigga.

At the end of the second week, her recovery seemed almost complete. The feathers she had so desperately clawed away were growing back along her tiny neck; her strength had astonishingly returned; and she had regained her muscular control, to the extent that she herself was obviously impatient to be gone. She no longer found simple pleasure in exercise; she flew direct for the windows, there to beat her little wings against the glass. Like every hospital patient, human or otherwise, whom we have ever known, she could hardly wait to be dismissed. With some misgivings, we took her outside and watched her take off and disappear into the trees. The male had not reappeared. We peered in vain the next few days through the binoculars at every female broadtail hummingbird who came near.

In the early evening hour of the fifth day after Bee Bop's departure, we were sitting outside near the back feeder when a hummingbird came flying unusually low over our heads. As it lit gently on a perch of the pine tree only a few feet away, we could identify it as a female broadtail. She flared her ragged tail, whose feathers had some shafts broken and a few quills separated. One wing, the right one, drooped ever so slightly. Bee Bop? After a moment or two of rest, she made a deliberate take off, flying just a little slower, and a little lower than our hummingbirds usually do, cleared the wood pile, and vanished again into the trees beyond.

So we entered in our "bird record," what was already engraved in our hearts, the Case History of Bee Bop:

June 8, 1966: Patient admitted. Diagnosis: Injury to muscular coordination; loss of locomotion. Prognosis: Unknown—doubtful recovery.

June 20, 1966: Patient dismissed; left under her own power.

June 25, 1966: Outpatient status. Two minute confirmation of restored health.

Gobi

Chapter Fourteen

GOBI

By a circuitous default we gathered Gobi into our family. We had no notion at all what we were letting ourselves in for, but faced with the decision today we would do it all over again. To be honest, Gobi has been very much more a plus than a minus in our house, though she came to us on a skimpy recommendation.

One of our World Family members who had come from India married a charming American girl and settled permanently in the United States. They decided they would embark on the adventure of owning a pet; and chose to begin with an Indian Hill Mynah bird whom they purchased from a pet shop in Denver. They named it Gobi in honor of Koshi's home continent, though certainly not in recognition of Gobi's home habitat in the hill country of Assam! But the three month old Mynah turned out to be incompatible with their household. Gobi was too young to talk, or even to try; she was extremely messy, both as to droppings and throwing her food about. Pat and Koshi worked and Gobi was left alone all day, a situation Mynahs particularly dislike, and under which they cannot thrive.

So Koshi called one evening in July of 1963 to see if we would like to adopt Gobi. As we had at that point two other birds, Mimer and Wigga, plus Kirk, all we needed was a third bird. But we went over to their house, "just to have a look," and returned home with an enormous parrot cage, a book about Mynah birds, some Mynah food pellets, and a seven-inch-long black fury enclosed in a 7½-inch box. So began the saga of "Life with Gobi," which is still going on some fifteen years later. It has been an experience of mutual education, with a generous share of delights and near disasters on both sides.

Life at 2049 the summer of 1963 was not in its calmest days when Gobi entered the ranks, for we were getting ready to embark on the great adventure of building Dunyana, and we were already psychologically, if not yet physically, camping in the place that had been home for eighteen years. Thus, arrangements for accommodating the new member of our household were various and temporary makeshifts, not conducive to her sense of security. There was literally no flat or empty place big enough to set her cage on except our breakfast room table. So we picnicked around the edges, as it were, and never had a real, sit-down-at-table meal from the time she joined us until some nine months later, when we moved from the notorious postage-stamp cabin to our completed Dunyana. She was a wild, nervously highstrung youngster, who squawked and rasped and snapped and flopped about in her cage if things frightened or startled her, which most things seemed to do a great deal of the time. Her bony structure was mature, but her plumage was not; she had hardly yet gotten her adult feathers, and her tail was perhaps the most humorous of her total, gangling, ill-fitting ensemble. It had two feathers, and when she brought her tail out at a right angle to her side for preening, Dale said it reminded him of the tiny illuminated excuse of an arm that used to shoot out from the middle of the left side of old VW's of pre-1960 vintage, as a turn signal. So, Gobi's tail, even in its adult, full 12-feather, squared-off form, is still her "turn signal."

Her enormously complicated, extra long, scaly, pale yellow feet with their black claws have changed very little, if any. Nor has her generous bill, which friends have likened to a giant piece of Halloween candy corn, dark orange shading to corn yellow near the tip. She has beautiful eyes, as do most other animal forms. They are large, very large, liquid brown, with the sparkle of a high degree of sharp intelligence shining in and through them.

Because she is extremely intelligent, Gobi is a complex, demanding personality. She is also willful and cantankerous. We were thankful we had amassed over the previous twelve years or so, a considerable backlog of experience with birds in the house, before taking on the project of living with Gobi. We could be more relaxed in dealing with her fears, her varied, physical needs, her pugilistic nature evidenced by the figurative chip she has always carried on her shiny shoulder. Maybe we were rather like the parents of a large family, who are undismayed by, and almost casual in handling, difficulties presented by the youngest child, difficulties which would drive a set of brand new parents right up the wall.

We would not, for instance, have been so sanguine about taking Gobi, almost immediately after acquiring her, to the cabin where we spent four of the most rugged months of the year. Not that she took the transplant serenely. Such was not her nature. During the memorable trip to the cabin, enclosed in her tiny box with a head-size hole for air, she rasped her disapproval and snapped at us if we leaned close to speak to her. I would have quailed at the thought (had she been our first bird) of subjecting a very young, tropical jungle bird, as yet barely acclimated, to her first winter spent in the high Rocky Mountains, in a house which had no heat at all at night, and where we all battled snow and cold a good part of the time! The instruction booklet said Mynahs should not be kept in temperatures lower than 65 degrees Fahrenheit, should never be left in a draught, nor allowed to get too hot. But by the time Gobi joined us, I had come to the more comfortable belief that most life is endowed with a measure of resistance and resiliency, and that with modest care and provision Gobi should be able to make it.

I did keep the cabin warm with three electric heaters during the day, and since it was so small it did not take very much to heat the rooms. The other two birds, Mimer and Wigga, were both hardy and well acclimated. Gobi proved to be stout as well, and even seemed to accept with assurance her privileged position atop the kitchen table. The nights were undeniably cold, however, so I tried to bridge that gap by purchasing an electric heating pad which we put over the top of her cage at night. This kept the temperature at a respectable degree for her; and she apparently came through our adventures there with less discomfort than any of the other animals—or I. Gobi can bear expert witness to the contention that Mynahs are "hardy" birds.

Gobi's cage came under our scrutiny before long. Since it was designed for a parrot not a Mynah, the one perch across the middle was too large in diameter for her to grab easily, and moreover too far from her feed dish for her to eat comfortably. The swing which hung from the middle, supposedly for entertainment, she regarded with aversion. So Laney took out the swing entirely, made some new perches in varying diameters and secured them at different angles. This redecoration seemed to meet with Gobi's approval, as she immediately hopped from one to another. She now has five perches, all of which she uses. Her cage has two bottoms, the upper one being a removable rack through which the droppings are supposed to fall and be caught by the solid bottom beneath. The bars of the rack are so far apart, however, that when she walks on them her feet sag in the middle. This looked extremely uncomfortable to us, so we have always kept paper towelling over that rack in order to make it easier for her and also because I cannot bear to think of standing on iron bars all one's life.

Gobi's cage is not a "cage," and has not been since we finally moved to Dunyana. I do not approve of caging a bird. In fact, every animal who has shared our home has been given the maximum liberty we could allow, restricted only by bowel habits which controlled how freely it could be permitted to roam. But for any living creature to be forever behind bars is anathema to me. Within a few weeks at Dunyana, her door was open from the first thing in the morning, when we took off her black night cloth, until we covered her after saying good night. For her part, Gobi obviously and unmistakably regards the cage as her exclusive territory but in no way as a forced confinement. She is content to remain within that territory, spends time frequently standing in the open door, knowing full well she can come out under her own power. Because she does—when she is frightened, or when she is lonesome. And she is territorially possessive about her "house."

The book on Mynahs said that the birds did not begin to imitate until they were about six months old. If she had been born in March when Hill Mynahs nest, Gobi was only three months old when we brought her home. She produced a variety of squawks, but seemed not to respond to any effort of ours to make her repeat a sound. At six months, according to the book, we would recognize an attempt to begin talking, by her leaning well forward on her perch and saying "urk." Gobi being Gobi, did no such thing. She leaned backward, and said a gurgly "ark." We agreed on "hello" as the first word we would try to teach her, and we said it a hundred times a day, in every conceivable voice and pitch. She would observe us with calm indifference, and perhaps turn her back, or she might emit an "ark." At last, after garbling to herself unintelligible sounds seemingly unrelated to "hello," she suddenly came forth one day with a perfectly articulated greeting. And she proceeded to elaborate by saying it in all the ways we had employed. Much delighted with this initial success, we waited a few days for its entrenchment, and tried the next lesson, which we had decided would be "How are you?" She seemed again to take no notice of the endless repetitions of this new sentence. Then to our consternation "hello" became unintelligible; and we thought we must have done something vastly wrong to cause such a retrogression. But in a few more days "hello" returned, bright as ever, and along with it, "How are you?" Once unscrambled, she was able to reproduce both lessons in as clearly enunciated English as we could speak.

She followed the same procedure over "Is that right?": apparent indifference, scrambling, reassembling, and final perfection of sound and pronunciation. This pattern she has followed ever since.

Mynahs are or have been most popular for their amazing imitative abilities, and while Gobi has exhibited her fair share of this talent, I find other qualities equally as engrossing. Gobi's sometimes uncannily apt use of words and phrases is deeply interesting to me because it is one of the evidences of her participation in our mutual environment. It is her total personality, much enhanced by her choices in the use of human language that makes her one of the most stimulating companions with whom we have ever lived.

She is also one of the most irritating, tyrannical, irascible, seemingly ungrateful, and hands-off-untouchable members of our family. For all that, she succeeds in being irresistibly loveable. For us, there is an irony in Gobi's longevity. The least responsive to caressing or handling, she has been with us the longest of any of our household animals, and shows no particular signs of aging at fifteen. It is even more frustrating for us to know that there *are* Mynahs who do enjoy being handled. But Gobi has required all of her fifteen years to overcome a minimum of objection to familiarity. We finally came to the conclusion that she must have experienced some traumatic circumstances in her very early weeks. First she exhibited downright fear which lasted for months; then remonstrance against the barest attempts to touch. After a couple of years we were permitted a stroke of her foot, followed by her one moderately affectionate or permissive time of day. This was in the evening, when I put her to bed, and paralleled in a way Mimer's "mushy" time first thing in the morning. When I pull up her white sheet from the back of her cage to cover her (it is pinned there permanently to spare the impaling of her food or even an occasional, forcefully ejected dropping on my drape just behind), prior to the additional black night wrapping, she hops to the top corner-wise perch and leans over, holding her head forward to be stroked. This has been her only self-initiated "mushy" time, and it may vary from several minutes and a general petting (through the bars) of her head and breast and back, even ruffling through her neck feathers, to a half dozen short strokes on her velvety head, when she announces the end by a snap of her strong beak, turns, and takes up her night's perch elsewhere.

Gobi's feathers are almost microscopically small on top of her flat head and they look and feel like the sheerest velvet. She has a predominantly purplish iridescence around her neck and on her raincoat feathers, while her rump is just as preponderantly a greenish sheen. Her collar (wattles) and side facial skin patches are a light and bright yellow closely matching the yellow of her bill tip and her long scaly legs. Her soft liquid brown eyes bely what one might expect—a gentle and loving disposition! The white patches close to the tips of her primary wing feathers are usually hidden, except when she stretches, either by extending each wing singly to its fullest length over her outstretched leg and foot, or by lifting her wings simultaneously until she looks like a giant butterfly. Then the white portion becomes a dramatic pattern in the dusky black. It also reveals how thinly she is feathered, because along her flank is a good deal of bare skin.

She is not the most graceful bird we have ever had, in repose or flying, when she stands or sits. She is not as attractive as Pecky in repose, and tends to stand with her long legs spread somewhat apart, sort of arms akimbo. Nor is she as graceful by far as Wigga in flying, as hers is a frantic flapping. Even her walk is more of an oddly gaited, sideways hop. This assortment of physical characteristics somehow turns into an integral expression of Gobi's personality.

She is a meticulously clean bird—about herself that is. She could care less how much mess she makes in or outside her cage. A great part of each day, especially when she is moulting, is spent investigating the state of each feather and making corrections or alterations. No particle of food or dropping may be allowed to remain on her for a second. She flips her food unmercifully on window, refrigerator door, floor, table, chairs, drain board, fastening thereon—as far as five feet away—banana bits, toast, shreds of the inside of a squeezed lemon (one of her favorite morsels when I am baking), and of course her regular diet of moistened fruit meal. (We tried installing a plastic guard but she quickly dispatched it in shreds.) But none of these items above is ever on herself. Neither will she step on a dropping. If she does so inadvertently, she jabs at her foot viciously, her every movement suggesting utmost revulsion. If she has decorated her perch in front of her food or water cups, she refuses to step on the perch and will instead hop clumsily to the edge of the dish itself, and teetering thereon, will "agg-agg" at me until I notice and remove the dropping. Then she immediately resumes standing on the cleansed perch.

She is very modest about her baths and requires complete privacy, by having her white sheet pulled all around the cage before she will get in, unless there is a terrific urgency because she is soiled. But I appreciate her requirements because it saves the entire kitchen from looking like a sauna room.

One of the irritants in life with Gobi is her undependability about baths. Sometimes we get our signals crossed and though she has asked for one, by clinging to her water dish, even stepping into it and saying "c'mon," a bath is not what she had in mind. Instead she uses her privacy inside the sheet to pull her towelling into the bath dish—but only partially. This results in capillary action emptying the bath dish and depositing a lake on the bottom tray. It is almost impossible to pull out that eighteen inch square tray with water close to its inch high rim, and carry it to the sink without spilling. Therefore I am annoyed, and I usually scold, telling Gobi she is a naughty bird.

She replies, promptly, and incisively, "Aw dry up!" Then in the most innocent of tones, as I clean up her mess plus the one I have made, "What'sa matter?"

Gobi *is* fascinatingly intelligent, so much a part of her several environments, so responsive to attention in her eagerness to participate that one comes to love and respect her in spite of her prickly disposition—maybe because she is so totally and honestly herself and does not try to dissimulate in order to please others. Almost all animals are honest and this priceless quality, if nothing else, makes them loveable. Gobi is nothing if not completely and disarmingly honest.

I am often contrite when I realize Gobi's mental potential and know that we did not persistently teach her to say all the words, phrases and sentences she could have mastered. She was apparently so capable of learning more, perhaps even craving to know more, that she has taught herself fully as many words and phrases as we have purposely taught her. Because of her sharp and intense powers of observation she has learned these without the amount of daily, accented repetition which we employed in our teaching. The fact that she learns to this day new sound effects convinces me she could learn new words, even if we did have to work longer to get results. I do not believe she is unable to learn, any more than I believe the human species is unable to learn at whatever age.

Perhaps her most remarkable self-taught sentence is "c'mon." It obviously is not one of her longer ones. But it assuredly involves a mental process.

Gobi was established in front of the glass wall of the kitchen at Dunyana from the time the workmen finished and we took over the house. It is a happy location for her. As most of the sustained inside activity takes place there and she can see over the whole hill outside, she is thus in the middle of everything.

It was next to her cage, in front of the drain board and sink area that I used to hold Kirk's dinner before him and tell him to say "please." As most dogs do, Kirk enjoyed prolonging the attention he got by delaying his "woof," and as a rule he was not so ravenously hungry that he could not wait. So he played his little game; and I would say, "Kirk, say 'please,' come on, come on." Apparently I repeated "c'mon" much more frequently than I said "please." Eventually Kirk answered and he was given his food. Gobi from her cage just above was taking all this in, and she observed that when I said "c'mon" Kirk finally said "woof," which must be "c'mon." But she herself had to make the connection that "c'mon" produced "woof" which produced food.

Before long, when I had a tidbit I thought Gobi would like I would say, "Would you like this, Gobi?" She began to answer "c'mon," and expanded to using it on her own initiative to ask for food. What seems amazing to me is that she drew a conclusion from what she saw and heard, and applied it to a successful gratification of her own needs and wishes. She now routinely asks in this way for something she wants. If it is a bath, she hops into her water dish, looks over at me and says "c'mon." If she wants a bite of grapefruit, strawberry, mushroom, hamburger, cheese, or whatever, out of her omnivorous appetite's choices, she hops back and forth from perch to perch, and repeats "c'mon." Just before I give it to her, I ask, "What do you say, Gobi?" She replies, "c'mon."

Another example of Gobi's use of language by association is her self-taught "g'bye." She uses it selectively, appropriately, never on command, but by interpretation of a situation. How she taught herself, or when, is a mystery, and like "c'mon," it is not a word said many times a day. It is one of her decisions, another means by which to participate in the life that goes on around her. No one ever asked her to say "g'bye." We had not yet gotten around to making it an assignment in the growth of her vocabulary, and we would have taught her to say "goodbye" anyway, not used the diminutive form.

As in learning to use "c'mon," she began by putting together elements of a routine. During the first years at Dunyana, Laney worked in Boulder. Upon finishing breakfast in the kitchen, she left the room by the exit to the den. A few minutes later she went downstairs to the front door. From the time she left the kitchen, Gobi could not see what was going on, but had to form her picture by ear. At some point between Laney's leaving the kitchen and her opening the front door we always said goodbye to each other. But Gobi began to anticipate the whole scene, and to say "g'bye" when Laney rose from the kitchen table. From that time on she has interpreted rising from the table and leaving the kitchen by the east exit, as the appropriate timing for a series of "g'byes."

She has added other signals which require "g'bye." When I bring my car keys to the kitchen table she has learned I am going to leave sooner or later, and she says "g'bye" as I put them down. She says it to me numerous times as I go downstairs. I find this poignant because she knows she is about to be alone and bored. When I am sewing in the next room, out of her sight, the sound of laying my scissors on the table resembles the keys, and Gobi pipes up, "g'bye."

She has her own "thing" going with our Brazilian member of the World Family, a student who is currently occupying the downstairs guest bedroom. If I am in another part of the house, I always know when Ricardo has come to the kitchen to prepare his breakfast because Gobi begins her conversation with him; that it is a varied monologue of her own repertoire does not deter her. It does not include "c'mon" because he is usually preoccupied and does not think to share his food with her. One morning, however, she was making such a racket that I came to see what was wrong. She was jumping back and forth on her perches, her eyes glued to the avocado which Ricardo was eating, and saying repeatedly, "c'mon." He looked puzzled and remarked, "I don't know what's wrong with Gobi. Something seems to be bothering her."

"She wants some of your avocado. That's her trouble," I replied. He immediately cut off a generous bite, which I diced finely in her tidbit dish. I had to push her away from the open cage door over which she was leaning in her eagerness, in order to put the dish in the corner. She ate it with such obvious relish we both laughed. That was the end of the commotion. She went back to her perch and settled down to digest her treat.

Though she has not previously included it during her one-sided conversation with Ricardo, she says "g'bye" without fail when he rises from the table, where he may have remained for an hour or more to read, after finishing his breakfast. She listens later for his step, from minutes to as long as a half hour, when he walks down the hall from his bedroom to the front door, and again, without fail, calls "g'bye" as he opens the door.

Answering the door bell, which requires opening and shutting the downstairs door, significantly, does not bring forth "g'bye," but "Hi!" "G'bye" is leaving, not coming, even though she cannot see what is going on, and must judge her comment by what her ears tell her.

She also helps out on telephone conversations with a running commentary which may vary in its content. Sometimes it may be embarrassing, as for instance, the time Laney was talking to a government official, and Gobi in the background kept saying "rat fink" and "aw get up," plus other derogatory and uncomplimentary things. The official sounded somewhat puzzled and a few strange silences occurred at the other end of the line. Gobi seems able also to judge when we are about to terminate a telephone call. Because seconds before we do so, she calls out "g'bye!"

The winter of 1971 I first took Gobi with me for a visit to Laney's home in Wyoming. She settled in promptly as a participating and contributing member of the household, quickly acquainted herself with the various signals of activity and assessed their meaning. Her post at Laney's is within view of most of what happens, as well as of both outside doors. She early decided that our going out the back door to feed the birds or to walk to the river is not actually leaving and she does not say "g'bye." But going out the front door is something else, applying to us as well as to guests. We found to our amazement to what extent she connected her perceptions and vocalized them. She soon read distant signs that we were going to leave the house. Laney kept our ski jackets with their rustling nylon exteriors in a closet out of Gobi's range of vision. Gobi had learned that our wearing those jackets meant that we were going away. She had also learned, obviously, the previous sounds accompanying our putting them on. We tried to test the extent of her conceptual picture, by putting the jackets on at the closet where she could not see us at all. But we had no sooner slid the door open, merely taking the jackets off their hangers then she began to call "g'bye!"

Recently, while Gobi and I were staying with Laney, we told a visitor that she interpreted the donning of a coat as "goodbye." He looked skeptical even though he is an admirer of Gobi's general intelligence. He had laid his coat on Laney's typewriter table at the other end of the room from her cage. When he picked it up, preparatory to leaving she promptly said "g'bye." He shook his head, and left, pondering her abilities. It has been said by some linguists that there can be no meaning nor symbolizing without the joining of two or more perceptions. Gobi's use of "c'mon" and "g'bye" seem clearly to be examples of this mental operation.

I don't know if Mynahs in their native habitat are accustomed to using water as a means to soften an otherwise harsh bite of food, but Gobi has learned on her own, to carry a bite of toast, a piece of cookie, which she finds too hard or dry for her liking, to her water dish. There she dunks it, takes it out and eats it, or she may drop it in the water and leave it until she considers it is "done," when she daintily picks it out. The only other time she is dainty about anything is when she does not really want a tidbit offered; then she very deliberately accepts it with the tip of her bill and just as deliberately drops it. Thanks but no thanks.

It is appropriate considering Gobi's general disposition that she cannot be bothered about saying "thank you" with the consistency of which she is capable. She does say it, however, beautifully and clearly, and in two distinct ways. One is an exaggerated "aa-n-k you-o-o," a perversion she considers we were guilty of when teaching her. The other is an entirely natural "thank you," as though she were broadly hinting that the ultra, prolonged effort was entirely unnecessary to impress the sound on her. She will repeat it as much on command as she repeats anything we ask, which is seldom. She says what she wants to say, when she wants to say it. But she will, sometimes, respond to either a compliment or a derogatory remark addressed to her with a quick "thank you" that is extremely apt and funny. If we laugh she repeats, and looks quite pleased with herself.

Her use of "Is that right?" which was one of the early sentences we taught her, is almost always humorously appropriate. She has two chief inflections, either very sarcastic, "Is *that* right!" or mere inquiry, "Is that *right?*" Her most flippant use, following a respectful silence, comes almost without fail in the middle of a diatribe I have been delivering from one of my soap boxes. When I have paused for breath she pipes up, and in the general silence says loudly, "Is *that* right!" and everyone dissolves in laughter. She regards laughter very properly as the feedback on which she thrives, and since the train of my lecture has been broken, she proceeds to launch into whatever part of her repertoire occurs to her. I am not the only recipient of her saucy intervention. She plays no favorites and chimes in on the conversation with anyone who is earnestly espousing a subject, interpreting the tone of voice as an invitation to put in her opinion.

She is the most brilliant conversationalist at any of our parties. I have never had a function at Dunyana where I did not find most of the guests eventually in the kitchen paying court to Gobi. At these times she simply sparkles with wit and pleasant repartee. Her joy in being the center of attention is such a complete contrast to Mimer, whose basic shyness was never overcome. He was always dismayed under the very circumstances on which Gobi thrives. She does not know what shyness is, and is happily extrovert, which is a great asset in winning people. She receives inquiries about her health from all over the world, as our World Family members have come to know and enjoy her during visits with us.

Gobi's chief delight when she goes to stay at my friend Margaret's (Gobi's "other mother" with whom she feels entirely at home when I go away) is the constant stream of people and activity of which she considers herself the center. Since she is in full view of everything, Gobi does not experience the frustration she has at Dunyana, where people have to make a trip to the kitchen to see her.

During one of Gobi's interludes at Margaret's she was present at a particularly notable dinner party. A retired Army Colonel, his wife, and some other guests were still at the dinner table which is within a few feet of the bay window where Gobi's cage stands at Margaret's house. The Colonel launched into one of his stories, that by consensus are regarded as rather long-winded affairs. He paused at a strategic point to take a breath, and Gobi who had been listening quietly filled the silence with "Is *that* right!" in her most sarcastic and disbelieving tone. In the ensuing laughter that burst forth, the Colonel's wife leaned over to Margaret and whispered, "I've been wanting to say that for a long time." But the Colonel saw nothing funny about it, and in distinct irritation turned around to Gobi and said, "Oh, shut up, bird." Gobi, quite unruffled, responded immediately. "Yes, sir!" she replied in clipped, military accents, just as she had been taught by Dale when he was home once on leave during his army stint. The Colonel has not regarded Gobi kindly since.

Margaret is much more able to laugh at herself, and has told with relish Gobi's retort that convulsed other friends who were visiting her one morning. Gobi had twisted to her own preferences a couple of phrases which became familiar to her early in our life together. During our first weeks at Dunyana, the plumbers were still about and they took great joy in teaching Gobi the wolf whistle (which she also uses with uncanny accuracy), and the then popular appellation, "rat fink." We were trying at the same time, to teach her to continue a conversation. Following "How are you?" which she says with great earnestness and great frequency, we wanted her to learn "I'm fine." However, what came out in the final product is perhaps more humorous and probably more accurate than the straight line, "I'm fine." For Gobi evidently considered "fine" and "fink" so similar it didn't matter which you used, so "I'm fine," became "I'm a rat fink."

On this particular morning when Margaret was showing off Gobi's language proficiency, she wanted her to say "I'm a rat fink." Usually quite cooperative on this matter, Gobi decided to be reluctant. Margaret repeated the phrase for her, several times, "I'm a rat fink, I'm a rat fink." Finally, after Margaret had announced she was a rat fink sufficiently clearly, Gobi answered her with "Is *that* right!" And the guests probably remembered Gobi's conversational ability much longer because of her embarrassing answer, than if she had dutifully repeated Margaret's words.

By slow, very slow degrees Gobi indicated to us that in addition to the attention she obviously enjoyed, she would tolerate a modest amount of affection at times other than her one voluntary "mushy" period. The days when Mimer was in the kitchen and we played with him, the petting Kirk received within her vision, even the rough-and-tumbles with Wigga, had not seemed to instill in Gobi any need to have the same for herself. Her reactions to both Kirk and Mimer were indeed more in the area of our conversations with them than with the animals themselves, perhaps because she was elevated off the floor and they were simply not on the same plane. She was removed physically from contact with them. (Except if Kirk got within snapping distance, when he put his nose up to the corner of her cage.)

With the coming of Bun Bun, our short-lived cottontail, and his nocturnal habits that nightly jarred Gobi, and especially with our feeding him on the drainboard right beside her cage during his illness, she did seem to indicate a dawning of pleasure at being stroked. Occasionally, I could reach through the bars and she would allow me to touch a foot, after which she inspected it carefully to make sure all the toes were still intact, a monitoring check she still observes.

Her affection began this way, then she asked for her nightly reassurance. She has slowly progressed to allowing me at almost any time to put my hand in the cage to stroke her breast, even to hold one foot in my hand.

Very recently when she has spent some time hopping about outside her cage, she seems not to mind our catching her, holding her close, and stroking her, when she becomes almost comatose. Small contented grunts she gives; then of a sudden, she comes to, almost panics to think what liberties she has permitted—and promptly bites. It is almost as though her wish to be loved, her very natural need for the precious emotional comfort and feeling of security which affection brings, runs counter to some early, traumatic lesson which warned her forever against trusting anyone. Only in the last year has she settled into my hand, stretched her neck to lay her head on my arm, and remained there, purring perceptibly, for a few moments of stroking, a truly amazing and rewarding development!

If Gobi has been reluctant to accept our affection she has shown even greater indifference toward any sort of relationship with other animals, except on a level of hostility. We had hoped that the position of her cage in the kitchen at Dunyana would encourage her interaction with life going on both inside and out. But the variety and almost constant presence of wild birds on the hill has not resulted in her learning their languages, even those of the year-round mountain chickadees or the Stellar's jays. While one of her calls does resemble a jay's that is as far as she has gone. She does seem to understand the jays' harsh danger signal when they streak across the hill scattering all the birds before them. She pulls in her feathers, does a bit of screaming herself, and frequently flies out of her cage. The call of fright must be the same in all languages. But otherwise, even on the balcony where the chickadees and siskins have sometimes come very close while we are sitting with her, she makes no attempt to talk with them nor do they do so with her.

Male and female Mynahs are not distinguishable by their

plumage. But we laughingly observed that one of the early indications of Gobi's sex was her declared and open preference for men. A man, even a man strange to her, can get away with the most outrageous familiarities, which, if a woman or a girl attempted them, she would reward with a sharp snap. Gobi is supposed to be a "soft-billed" bird, but there is nothing soft about her bill; her bite leaves a slash as neat as a paper cut, and even more painful because it is a bruise in addition. But she has yet to bite a man for putting his hand in her cage and stroking her. Some men have succeeded in getting her to stand on their hands. She will take their proffered food in the gentlest and most well-bred fashion, but let a woman, and most especially, Laney, offer it, and Gobi becomes a brazen, pugnacious little hussy, and snaps at her fingers before even acknowledging the food.

As might be expected, her behavior toward men charms and captivates them. One time we entertained as our house guests a very proper, handsome, cool and collected diplomat and his wife from the Middle East; he was the very essence of the polished, worldly gentleman. He and Gobi took to each other immediately, especially since Gobi by then had learned her first foreign language sentence, the Arabic "ahlan wa sahlan" greeting, which she spoke impeccably. His wife, however, got the usual treatment meted out to women. When we were gathered at the front door downstairs ready to depart for the airport, Rashid suddenly said, "I must go back and say goodbye to Gobi once more." So he climbed the stairs to pay his respects to our little black termagant!

There is a qualitative differential in Gobi's treatment of Laney, between her actions here at home and her behavior at Laney's house. As clearly as deeds can speak, Gobi shows that at Dunyana she feels Laney gets attention from me that should be Gobi's. And Laney being nothing loath to challenge Gobi, the fight is on almost as soon as Laney comes in the front door. Recently she bit Laney as she offered her a morsel of food from the side of the cage instead of at the open door. Provoked, Laney moved to the door where she has frequently reached in and grabbed her, under similar provocation. As she came to the door, but without her hands out to catch her, Gobi fled anyway, to the far corner perch, and sat there looking as though she knew she had invited retaliation. Yet ten minutes later she was charming a couple of male utility employees who had come to fix one of my ovens, by pleasantly allowing them, perfect strangers, to put their hands in her cage and stroke her.

However, when I bring Gobi to Laney's (an undertaking not unlike travelling with a baby and all its ponderous modern gear, including play pen), her deportment is mild, even verging on the acceptable. Laney can feed her by hand at her house without Gobi's being more interested in biting her first and taking the food second. She seems to realize that she is on Laney's territory where she is not top dog. She is also much more polite and friendly to Margaret at her house than when Margaret comes to Dunyana to see us. Without any doubt she is the prickliest cockleburr we have ever had to live with; not even excepting Emmy who was not the most obliging character I have known, and whose personality Gobi most resembles.

Living in the kitchen as she does, Gobi has inevitably become an expert on foods. Nothing going in or coming out of the refrigerator, the freezer, or the oven escapes her sharp and thoughtful inspection. Nor does she consider any food out of bounds, at least for experimentation, except liquor.

My unpacking the grocery bag is a highlight of her day. With uncanny clairvoyance, it would seem, she spots a package of ground round wrapped in plastic, sees the mushrooms in their box, the strawberries, broccoli whose flowerets she loves, spinach, and blueberries which she can inhale in surprising numbers; without moving her feet her whole body follows the food across the kitchen to the refrigerator. Later when I take them from the refrigerator to the sink, she follows them back again, knows it is now time for her to be given some and begins to beg. She has some sort of computer memory, as did Mimer, about how much of any certain delicacy still remains in the refrigerator. Begging consists of short little half grunts, and a constant hopping from one perch to another, quite a sufficient disturbance to catch anyone's attention. All bananas she considers quite logically belong to her. Her first food each morning is a half-inch slice of banana, cut into tiny squares in her tidbit dish and placed on the bottom of her cage in a corner. If she has not already said "c'mon" while I fixed it she does so as I hold it in front of her. She knows she will not get any tidbit until she says her please. Sometimes she has several other remarks to make before getting down to food, but eventually she gets around to "c'mon."

Gobi absorbed and added to her vocabulary a question which I used to ask Mimer during inclement days when he had to be in the kitchen. He would become so restive I would ask him "What's the matter, Mimer?", to which he always replied volubly. Out of that interaction which took place close to her cage, Gobi decided to extract, "What'sa matter," which she accurately applies when I am doing one of the five or six daily sweep-ups under and around her cage. She comes down to the open door, peers over and asks her question in a very concerned tone. More noteworthy is her ability to sense if I am distracted or emotionally upset, when she again inquires solicitously, "What'sa matter?"

The marvelous faculty of sound imitation, in part a defense mechanism in the Mynah's natural habitat, Gobi uses for expert communication with her human surroundings and on another level than that of survival. How fascinating it would be to observe how the wild Mynah uses its particularly complex aid to survival for communication socially with its own kind. The clarity of Gobi's speech is one of her most remarkable capabilities: without a larynx, having to "cough up" her words, without lips or teeth, to be able to form "F," "B," "TH," "M" as articulately as we!

The plumbers (having been responsible for "rat fink"), and as an additional contribution to Gobi's vocabulary, taught her with glee the wolf whistle. I cannot whistle, so she does not hear it very often. When she does use it there always seems to be an hilariously funny and appropriate reason.

One summer day when the doors and windows were open, a car came up the road, around the cul de sac, and parked above our house. Gobi announced its arrival as usual and kept a watchful and anxious eye on what went on up there, stretching her neck and changing her position to get a better look. The driver got out and opened the hood of his car, where there was apparently some trouble. He was leaning way over it, when Gobi brought forth her whistle, a low and vivid one. Unmistakable! The man drew his head out from under the hood, turned around and looked inquiringly at the house. He could see no one, and Laney and I, after Gobi's call, did not think it wise to appear. So he went back to his investigation. Again Gobi whistled. Again he looked around, a little irritated this time. It was now obviously embarrassing for us to attempt any explanation. Furthermore, we were now convulsed with laughter. Again he went back to his engine trouble. And once more Gobi gave a long, deliberate, piercing wolf whistle. This time the man drew back from the hood, slammed it shut, got in the car and drove off furiously in a cloud of dust. Gobi enjoyed our feedback and probably considered her efforts a success.

Gobi increasingly demonstrates that in addition to her use of language she acts on the basis of concepts (pictures) or ideas. These mental attributes are clearly possessed by other forms of life. Why it is so difficult for mankind to acknowledge these abilities I cannot understand. What beauty we deny ourselves by refusing to see the world through that widened lens of appreciation. We would be free to recognize and share parallel qualities, abilities, depths in our fellow creatures. It would strip off the blinders that prevent our blending into the marvelous workings of the universe of which we are a part, and rid us of the obsession to conquer instead of cooperate.

We would find as an end in itself the joy of seeing and understanding what goes on about us: the loving provision of food a chickadee bestows on his mate when she is nesting; the sensuous luxury a cottontail enjoys stretching his full length on the springy pine needles in the warmth of a summer morning; the sheer enjoyment of the warm standstone one of my eastern fence lizards feels when on a cool day he hugs the rock with his whole underside. What we humans most need is to bring ourselves into creative alignment with the rest of Nature, to make ourselves worthy to belong in the great scheme of things. This requires a humility, a true humility not a false "meek-humble" humbug, a real willingness to see the worth of other species for what they are in themselves, and to grant that they have a right to live for their own fulfillment, not only in the limited frame of their "usefulness" to man.

Gobi has her own sensuous creature comforts which she has learned to enjoy, and which she tries to achieve for herself. All through the winter she never begs to be taken out on the balcony for a sun bath which she so enjoys. But after the sun has arrived at a certain meridian, Gobi begins to jump around on her perches following breakfast. If I appear to be heading for the door to the balcony she "ags" and almost falls out of her cage in her eagerness to go along. This timing has to be a concept in her head, however we want to name it, of the proper combinations of light, warmth, and my actions. When put together these produce trips to the balcony and pleasant experiences. Sunbathing is a specialty she enjoys and indulges in according to the rules of her particular genes. These include spreading a wing, putting her head far to the side, raising her neck and head feathers so the sun can get to her pale skin. This without fail begets itching within a few moments, when she comes out of her glazed-eye stupor to scratch violently, then returning for more radiation.

Gobi carries in her little black velvet head images of our actions which have certain results. This is association. Maybe she always has observed, has always been aware, but she now makes known what seems to be a widening comprehension. For example, she first responded to my goodbye as I went out the kitchen door or downstairs to the garage. Then she said goodbye on her own when I put on a coat in her presence. Next she preceded that clue by saying goodbye when I put on my high boots to go out on the hill. Now there are increasing signals, which are farther and farther removed from my actual departure. Pulling the kitchen drapes is one, closing her cage door, setting car keys on the table, putting on gloves, even taking off my house shoes before going near the boots, setting out the pan of bird seed—it is daily fun and stimulation to observe and participate in Gobi's understanding of her surroundings.

Another example of her increased number of signposts has surfaced over the cocktail hour. She has become an enthusiastic habituée along with us. Though she disdains all alcoholic beverages, she knows that cheese frequently accompanies our drink, and this she loves with a passion, any and all kinds! She begins her "c'mon" now with the opening of the liquor cabinet, anticipating the remaining preparations which produce her share of the goodies.

There are for her some signs which must lie deep in her genes. Danger is imminent to her in big, tube-like objects, and she has a "tizzy" if a rolled tablecloth or a mailing tube larger than a couple of inches in diameter comes within sight. She flops around her cage, screeching in terror. Yet the hose of my vacuum cleaner does not alarm her at all. Could she be reminded of a python or a boa constrictor, something she has never actually experienced? What else but a racial knowledge could trigger a reaction so specific? Or her instantaneous panic when a hawk flies over our hill?

Looking at Gobi the other day I realized again in frustration that I can never capture the essence of her on reams of paper, no matter how detailed, how lovingly I try to describe her and her activities. Not even pictures, potent as they are, can depict the real shine in her big brown eyes; the way she can change her expression by altering their shape, along with the rise and fall of her head feathers; nor her scrappy belligerence alternating with appealing moments, her mercurial disposition. Gobi is my ongoing challenge, a creative companion, a rewarding if exasperating responsibility, whose influence flows from her tiny body to penetrate the whole house and my life.

It is a thwarting deficiency of language that it cannot incorporate all the dimensions of a life. Everything is once or twice removed, changed, and reduced in the expression. I have found it so in my efforts to bring alive the animals I have loved and lived with. I cannot make alive the rounded flesh, the breathing personalities. I can only set down in one-dimensional words my profound respect for their intelligence, for what they have taught me, singly and collectively, and record my gratitude for being loved by them.

THE END

Epilogue

The unassuming honesty, the integrity of character which each of my animals has possessed always touched me deeply. These friends truly earned my respect and love; and they taught me, further, to look upon members of my own species in the same way. I found long ago, as I reached out toward a wider world of human people, that other shades of skin, other languages and cultures vastly different from my own were no barrier at all to appreciation of our mutual humanity.

The animals and I have always lived together on the basic premise that only our native endowments were important. While human association is a much more complex affair, if one sheds the assumptions of superiority, or the chauvinism of provinciality, the deeper wellsprings of kinship and that great leveller, a sense of humor, are set free. Thus has my World Family grown and the bonds strengthened, now into the second generation.

With training from those "other nations," I have become a comfortably liberated citizen of the world. My loyalties, my chauvinism, are on behalf of our lovely planet and its diversity of life and beauty, so delicately bound together, and yet so indissolubly linked. All this I learned by example, and in the beginning, from my animals.

I hope the people who read this book will be joyfully reminded of their own friends in fur and feathers, with whom they have shared precious years. Maybe they too will recognize that these clear-eyed, forthright creatures have given them the ability to look at the foibles of our human world through a little wider lens, with a little more clarity.

Chronology

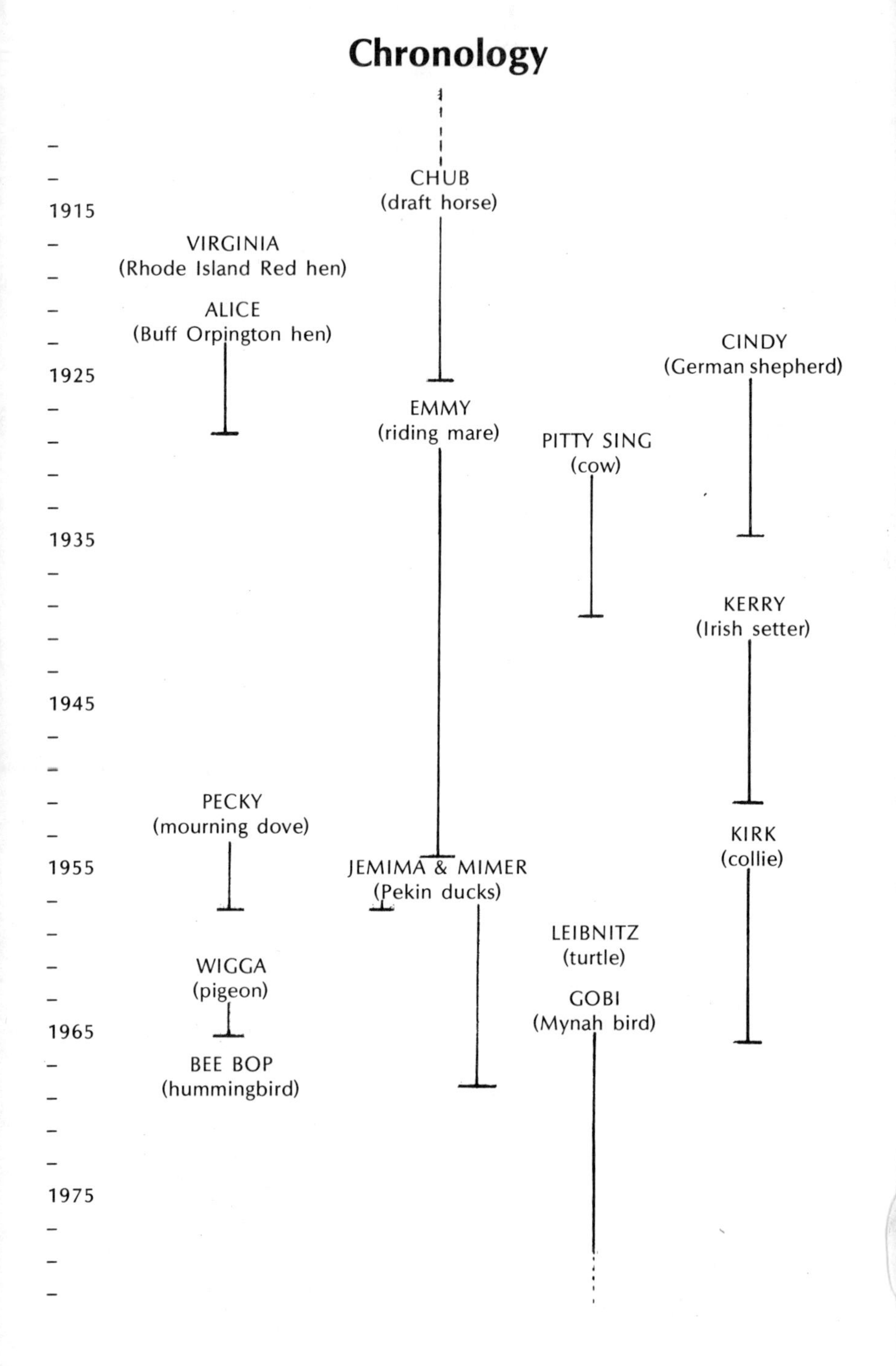

1915
1925
1935
1945
1955
1965
1975
CHUB
(draft horse)
VIRGINIA
(Rhode Island Red hen)
ALICE
(Buff Orpington hen)
CINDY
(German shepherd)
EMMY
(riding mare)
PITTY SING
(cow)
KERRY
(Irish setter)
PECKY
(mourning dove)
KIRK
(collie)
JEMIMA & MIMER
(Pekin ducks)
LEIBNITZ
(turtle)
WIGGA
(pigeon)
GOBI
(Mynah bird)
BEE BOP
(hummingbird)